中国海洋统计年鉴

CHINA MARINE STATISTICAL YEARBOOK 2015

国 家 海 洋 局

Edited by

State Oceanic Administration,

People's Republic of China

海洋出版社

China Ocean Press

《中国海洋统计年鉴》编委会

马云翔　　中国船舶工业集团公司
刘　悦　　中国船舶重工集团公司
侯启军　　中国石油天然气集团公司
邓瀚深　　中国石油化工集团公司
金晓剑　　中国海洋石油总公司
刘苗夫　　中国盐业总公司
李志军　　辽宁省海洋与渔业厅
于忠信　　河北省国土资源厅
孙连友　　天津市海洋局
王伟杰　　山东省海洋与渔业厅
张建军　　江苏省海洋与渔业局
沈依云　　上海市海洋局
童加朝　　浙江省海洋与渔业局
廖信宁　　福建省海洋与渔业厅
洪伟东　　广东省海洋与渔业局
刘　斌　　广西壮族自治区海洋局
潘建纲　　海南省海洋与渔业厅
栾玉瑄　　大连市海洋与渔业局
傅道军　　青岛市海洋与渔业局
吴建义　　宁波市海洋与渔业局
王春生　　厦门市海洋与渔业局
梁俊乾　　深圳市海洋局

《中国海洋统计年鉴》编辑部

徐　婧　中国船舶重工集团公司
肖　勇　中国石油天然气集团公司
孙　艳　中国石油化工集团公司
郝静辉　中国海洋石油总公司
杜丽华　中国盐业总公司
魏　南　辽宁省海洋与渔业厅
程　新　河北省国土资源厅
杨金璐　天津市海洋局
张收元　山东省海洋与渔业厅
顾云娟　江苏省海洋与渔业信息中心
张　呈　上海市海洋事务管理中心
王志勇　浙江省海洋与渔业局
廖润雪　福建省海洋与渔业厅
利红兵　广东省海洋与渔业局
曹淑萍　广西壮族自治区海洋局
薛羽捷　海南省海洋与渔业厅
申蓉连　大连市海洋与渔业局
肖本统　青岛市海洋与渔业局
吴鸣春　宁波市海洋与渔业局
池信才　厦门市海洋与渔业局
罗小霞　深圳市经济贸易和信息化委员会

执行编辑：李长如

英文校订：林宝法

Zhang Jilin	National Tourism Administration
Shi Bing	Chinese Academy of Sciences
Han Zhiqiang	China Earthquake Administration
Xie Pu	China Meteorological Administration
Ma Yunxiang	China State Shipbuilding Corporation
Liu Yue	China Shipbuilding Industry Corporation
Hou Qijun	China National Petroleum Corporation
Deng Hanshen	China Petrochemical Group Corporation
Jin Xiaojian	China National Offshore Oil Corporation
Liu Miaofu	China National Salt Industry Corporation
Li Zhijun	Department of Ocean and Fisheries of Liaoning Province
Yu Zhongxin	Hebei Province Department of Land and Resources
Sun Lianyou	Tianjin Ocean Administration
Wang Weijie	Department of Oceanic and Fishery of Shandong Province
Zhang Jianjun	Jiangsu Provincial Ocean and Fisheries Bureau
Shen Yiyun	Shanghai Ocean Administration
Tong Jiachao	Zhejiang Provincial Ocean and Fisheries Bureau
Liao Xinning	Department of Oceanic and Fishery of Fujian Province
Hong Weidong	Guangdong Provincial Oceanic and Fishery Administration
Liu Bin	The Oceanic Administration of Guangxi
Pan Jiangang	Department of Marine and Fishery of Hainan Province
Luan Yuxuan	Dalian Ocean and Fishery Administration
Fu Daojun	Qingdao Ocean and Fishery Administration
Wu Jianyi	Ningbo Ocean and Fishery Bureau
Wang Chunsheng	Oceans and Fisheries Bureau of Xiamen
Liang Junqian	Shenzhen Ocean Administration

编 者 说 明

一、《中国海洋统计年鉴》2015年版是一部全面反映2014年中华人民共和国海洋经济发展和海洋管理服务情况的资料性年鉴，全书为中英文对照。

二、本年鉴的统计资料范围为人们在海洋和沿海地区开发、管理、利用海洋资源和空间，发展海洋经济的生产和活动以及沿海地区的社会经济概况。地域范围为沿海地区、沿海城市和沿海地带，其排列顺序按《沿海行政区域分类与代码》(HY/T 094—2006)的顺序排列。

三、本年鉴内容包括综合资料、海洋经济核算、主要海洋产业活动、主要海洋产业生产能力、涉海就业、海洋科学技术、海洋教育、海洋环境保护、海洋行政管理及公益服务、全国及沿海社会经济、部分世界海洋经济统计资料等十一部分。

四、本年鉴根据《海洋统计报表制度》（国统制[2012]50号）和《海洋生产总值核算制度》（国统制[2013]93号），资料主要来源于沿海省、自治区、直辖市统计局、海洋厅（局）以及20个有关涉海部、局、总公司。

五、本年鉴中除特殊说明外，所有价值量指标均为当年价，除注明年份外，其他均为2014年数据，年鉴中每部分后附有主要海洋统计指标解释，对主要海洋统计指标的含义、统计范围和统计方法做了简要说明。统计数据中的其他说明置于表的下面。有续表的资料，如有注释均置于最后一张表的下面。

六、由于第三次全国经济普查对2009—2013数据进行了修正，故本年鉴中“全国及沿海社会经济”和“海洋经济核算”中有关数据为修正后数据。

七、本年鉴表格中符号使用说明：“…”表示数据不足本表最小计算单位数；

“空格”表示该项统计指标数据不详或无该项数据；“#”表示其中的主要项；其他符号如“*”或“①”等表示本表后面有注释。

八、本年鉴中由于数字精确度的原因，四舍五入后部分分项值之和与合计值有微小差异。

九、本年鉴资料国内部分未包括香港特别行政区、澳门特别行政区和台湾省数据。

十、《中国海洋统计年鉴》在编撰过程中，得到了各有关单位的大力支持，我们在此表示衷心的感谢。本年鉴中如有疏漏和不妥之处，敬请读者批评指正。

《中国海洋统计年鉴》编辑部

Introduction

I. *China Marine Statistical Yearbook* (2015) is a data almanac which reflects in an all-round way the development of marine economy, marine management and service in the People's Republic of China in 2014, and it is a Chinese-English bilingual edition.

II. The Yearbook's statistics cover the production and activities in the marine and coastal areas in relation to the development, management and utilization of marine resources and space, and the development of marine socioeconomy. The regions covered are the coastal regions, coastal cities and coastal zones with coastlines, which are arranged in order according to the *Coastal Administrative Areas Classification and Codes* (HY/T 094-2006).

III. The data in the yearbook consist of 11 sections, namely, integrated data, marine economic accounting, major marine industrial activities, production capacity of major marine industries, ocean-related employment, marine science and technology, marine education, marine environmental protection, marine administration and public-good service, national and coastal socioeconomy, part of the world's marine economic statistics data.

IV. The Yearbook is based on the *Marine Statistics Report System* (Guotongzhi [2012] No. 50) and the *Ocean Gross Product Accounting System* (Guotongzhi [2013] No. 93), its data mainly come from the statistical bureaus and the oceanic administrations of the coastal provinces, autonomous regions, and municipalities directly under the Central Government as well as the 20 ocean-related ministries, bureaus and general corporations concerned.

V. Unless otherwise specified in the Yearbook, all the value indicators are given at the current price. Except for those noted in years, all the others are the data of 2013. Each section is attached by explanatory notes to the major marine statistical indicators, giving a brief explanation for the meaning, statistical range and statistical methods of the major marine statistical indicators. Other notes to the statistical data are listed below the tables. For the data with continued tables, annotations, if any, are put below the last table.

VI. As the Third National General Economic Survey has revised the data of 2009-2013, the related data in the "National and Coastal Scio economy" and "Marine Economic Accounting" of the Yearbook are the revised ones.

VII. The usage of symbols in the tables: "…" indicates the statistics smaller than the minimum calculation unit in the table; "Blank" indicates that the data of the statistical index is unknown for the time being or that there is not such data available; "#" indicates the major items of the table; Other symbols, such as "*" or "①", indicate "see footnotes below".

VIII. For reasons of digital accuracy, there is small difference between the sum of values of some subterms after having been rounded off and the total values.

IX. The domestic part of the Yearbook does not include that of Hong Kong Special Administrative Region, Macau Special Administrative Region and Taiwan Province.

X. In the course of editing *China Marine Statistical Yearbook*, we enjoyed energetic support from the various departments concerned and we hereby extend our heartfelt thanks to them. Criticisms and comments are welcome from readers on any of the oversights and inappropriateness as the time for editing is too short.

Editorial Department of
China Marine Statistical Yearbook

目　次
CONTENTS

3 主要海洋产业活动
Major Marine Industrial Activities

5 涉海就业

Ocean-Related Employment

6 海洋科学技术

Marine Science and Technology

7 海洋教育

Marine Education

8　海洋环境保护与防灾减灾
Marine Environmental Protection and Disaster Mitigation

9 海洋行政管理及公益服务
Marine Administration and Public-Good Service

11 部分世界海洋经济统计资料
Part of the World's Marine Economic Statistics Data

2014 年我国海洋经济发展综述

2014 年，各级海洋行政主管部门认真贯彻落实党中央、国务院建设海洋强国和 21 世纪海上丝绸之路的战略部署，海洋产业转型升级势头良好，海洋经济在新常态下保持平稳的发展态势。

一、全国海洋经济发展概况

2014年，全国海洋生产总值60 699亿元，比2013年增长7.9%（除特别注明外，增长率均按可比价计算），海洋生产总值占国内生产总值的9.54%，占沿海地区生产总值的16.3%。全国涉海就业人员3 554万人，比2013年增加39.4万人。

二、主要海洋产业发展情况

2014年，主要海洋产业实现增加值25 303亿元，比2013年增长8.1%，占海洋生产总值的41.7%，滨海旅游业和海洋交通运输业仍占主导地位。

海洋第一产业 2014年，海洋渔业实现持续稳定增长，海洋水产品产量加快增长，海水养殖及远洋渔业生产能力不断提高。海洋水产品产量为3 296.2万吨，比2013年增长5.0%。其中，海水养殖产量继续增加，达到1 812.7万吨，比2013年增长4.2%；海洋捕捞产量也有所提高，为1 280.8万吨，比2013年增长1.3%；由于远洋捕捞能力的加强，产量增长较大，达202.7万吨，比2013年增长50.0%。海水养殖面积略有下降，为230.6万公顷，比2013年减少0.4%。远洋渔船数量达到2 460艘，比2013年增加13.9%；总功率达到202.6万千瓦，比2013年增长27.9%。

海洋第二产业 2014年，受国际原油价格的影响，海洋油气业呈现下降，全年实现增加值1 530.4亿元，比2013年减少5.0%，海洋原油产量4 614万吨，比2013年增长1.6%，海洋天然气产量130.9亿立方米，比2013年增长11.3%。海洋矿产资源开采秩序进一步规范，全年实现增加值59.6亿元，比2013年增长11.2%。海洋盐业平稳发展，全年实现增加值68.3亿元，比2013年增长8.8%。海洋化工产业发展较快，全年实现增加值920亿元，比2013年增长15.4%。海洋生物医药产业持续稳步发展，全年实现增加值258.1亿元，比2013年增长7.3%。随着海上风电项目有序推进，海洋电力业发展较快，全年实现增加值107.7亿元，比2013年增长17.6%。海水利用业产业化水平不断提高，产业继续保持发展态势，全年实现增加值12.7亿元，比2013年增长3.7%。海洋船舶工业加快调整转型步伐，产业发展较快，全年实现增加值1 395.5亿元，比2013年增长15.5%。海洋工程建筑业继续保持稳步增长，全年实现增加值1 735亿元，比2013年增长7.5%。

海洋第三产业 2014年，全国沿海港口生产形势总体良好，但由于供求失衡加剧，国内航运市场仍在低迷与复苏之间徘徊，全年海洋交通运输业实现增加值5 337亿元，比2013年增长9.6%。沿海港口货物吞吐量803 307万吨，比2013年增长6.2%，国际标准集装箱吞吐量18 178亿标准箱，比2013年增长7.1%。滨海旅游产业规模继续扩大，沿海地区在加大旅游基础设施投入的同时，不断丰富海洋旅游产品，海洋邮轮游艇旅游发展步入新阶段。主要沿海城市接待入境旅游者人数3 886.6万人次，比2013年增加107.0万人次；滨海旅游业通过不断深挖国内旅游市场，全年实现增加值9 752.8亿元，比2013年增长11.1%。

三、区域海洋经济发展情况

2014年，环渤海、长三角、珠三角三大经济区加快结构调整步伐，海洋经济继续保持平稳增长的态势。环渤海经济区、长江三角洲经济区和珠江三角洲经济区海洋生产总值分别为22 288.9亿元、17 276.9亿元和13 229.8亿元，占全国海洋生产总值的比重分别为36.7%、28.5%、21.8%。

环渤海经济区海洋经济总体增速放缓，海洋生产总值比2013年增长7.6%（现价），占地区生产总值比重15.9%。海洋产业增加值为13 265亿元，海洋相关产业增加值为9 024亿元。海洋渔业、海洋油气业、海洋交通运输业、滨海旅游业依然是该区海洋经济的支柱产业，增加值合计达到8 773.7亿元，占该地区主要海洋产业增加值的84.8%。滨海旅游业和海洋电力业实现快速增长，分别比2013年增长23.4%和17.4%（现价）。

长江三角洲经济区实现增长态势，海洋生产总值比2013年增长10.9%（现价），海洋生产总值占地区生产总值比重13.4%。长江三角洲海洋产业增加值10 243.8亿元，海洋相关产业增加值7 033.1亿元。按照产值贡献，滨海旅游业、海洋交通运输业、海洋船舶工业和海洋渔业四个产业位居前列，其增加值之和占该地区主要海洋产业增加值的93.4%，其中滨海旅游业占该地区主要海洋产业增加值的42.6%，产值贡献位居第一。海洋船舶工业、海洋电力业、海洋工程建筑业增速较快，分别比2013年增长23.0%、23.0%和19.5%（现价）。

珠江三角洲经济区继续保持稳步增长，海洋生产总值达13 229.8亿元，比2013年增长13.1%（现价），海洋生产总值占地区生产总值比重达19.5%。海洋产业增加值为8 167.6亿元，海洋相关产业增加值为5 062.1

亿元。滨海旅游业、海洋交通运输业、海洋化工业和海洋油气业依然为珠江三角洲经济区海洋经济发展的支柱产业，其增加值之和占该地区主要海洋产业增加值的80.0%。

四、海洋科研教育

2014年，海洋科研教育事业继续保持稳步发展，统计的海洋科研机构共189个，从业人员40 539人，比2013年增长4.6%；海洋科研机构承担海洋科技课题17 702项，比2013年增长8.4%；发表海洋科技论文16 908篇，出版海洋科技著作314种；专利授权数4 020件，其中发明专利2 647件，比2013年增长17.9%。开设海洋专业的高等院校达408个，比上年增长3.8%，专任教师数306 669人，比上年增长2.5%。高等教育和中等职业教育海洋专业毕业生数82 631人，比上年下降12.6%；招生人数和在校人数分别为71 755人和233 059人，分别比上年下降3.8%和9.0%。

五、海洋环境保护与防灾减灾

2014年，我国海洋环境质量状况基本稳定，与2013年基本持平。海水的放射性水平和海洋大气γ辐射空气吸收剂量率均未见异常。国家级海洋保护区生态状况基本保持稳定。重点海水浴场、滨海旅游度假区环境状况总体良好。海水增养殖区环境质量状况基本满足养殖活动要求。海洋倾倒区环境状况总体稳定，未因倾倒活动产生明显影响，近岸海域平均富营养化状况为轻度富营养。但是，近岸海域海水污染、海洋生境退化、环境灾害多发等问题依然突出。陆源排污压力巨大，近岸局部海域污染依然严重；81.0%实施监测的河口、海湾等典型海洋生态系统处于亚健康和不健康状态；赤潮发现次数和累计面积与近5年平均值基本持平；渤海滨海平原地区海水入侵和土壤盐渍化严重；局部砂质岸段和粉砂淤泥质海岸侵蚀程度加大。2014 年，我国海洋灾害以风暴潮、海浪、海冰和赤潮灾害为主，绿潮、海岸侵蚀、海水入侵与土壤盐渍化、咸潮入侵等灾害也均有不同程度发生。各类海洋灾害造成直接经济损失136.14 亿元，死亡（含失踪）24 人。

六、海洋行政管理

2014年，行政管理工作扎实推进，各项海洋工作取得明显成效。全年颁发海域使用权证书5 011本，比2013年增长27.3%；确权海域面积374 148公顷，比2013年增长5.4%；全年共签发疏浚物海洋倾倒许可证375份，实施各项海洋执法检查共145 170次，发现违法行为1 832起；提供海洋数值预报服务共29 395次，开展海洋调查项目832个，获得数据共161.9万个，各项海洋观测获得数据量共32 253.5万个，全年接收存档卫星遥感数据量共计41 719.7GB；本年接收纸质档案21 640卷（册），电子档案3 170GB；共有18项国家标准和94项行业标准通过立项审查，发布国家标准17项，行业标准12项。

Summary of China's Marine Economic Development in 2014

In 2014, the competent marine administrative departments at all levels conscientiously put into effect the strategic plan of the Central Party Committee and the State Council for building a sea power and the maritime silk road, the momentum of transformation and upgrading for marine industries was good and the marine economy kept a stable posture of development under the new normal state.

I. Survey of the National Marine Economic Development

In 2014, the national gross ocean product amounted to 6 669.9 billion yuan, 7.9% up from that in 2013 (Unless otherwise specified, the growth rate is calculated at the comparable price), accounting for 9.54% of the GDP and 16.3% of the Gross Regional Product. The number of people employed by the marine-related sectors throughout the country reached 35.54 million, 0.394 million more than that in 2013.

II. Development of Major Marine Industries

The major marine industries realized an added value of 2 530.3 billion yuan in 2014, registering an increase of 8.1% as compared with 2013, accounting for 41.7% of the gross ocean product. Coastal tourism and marine communications and transport still occupied the leading position.

Primary marine industry

In 2014, the marine fishery kept a sustained and stable growth, the output of marine aquatic products quickened its pace of growing and the capacity of mariculture and ocean-going fishery is constantly being strengthened. The yield of marine aquatic products amounted to 32.962 million tons, 5.3% up from 2013, among which, the output of mariculture continued to grow, reaching 18.127 million tons, 4.2% up from 2013; the yield from marine fishing increased to some extent, amounting to 12.808 million tons, 1.3% up from 2013; thanks to the improvement of the ocean-going fishing capacity, the output witnessed a major growth, reaching 2.027 million tons, 50.0% up from 2013. The area of mariculture somewhat dropped, amounting to 2.306 million hm^2, decreasing by 0.4% as against that in 2013. The number of ocean-going fishing vessels reached 2 460, 13.9% up from 2013; the total power amounted to 2.026 million kW, 27.9% up from 2013.

Secondary marine industry

In 2014, affected by the price of international crude oil, the offshore oil and gas industry presented a posture of declining, effecting an added value of 153.04 billion yuan for the whole year, 5.0% down from 2013. The output of marine crude oil was 46.14 million tons, 1.6% up from 2013 and that of marine natural gas 13.09 billion m^3, growing by 11.3% as against that in

2013. The order of marine minerals mining had been further normalized and a full-year added value of 5.96 billion yuan had been effected, 11.2% up from 2013. The marine salt industry developed steadily, effecting a full-year added value of 6.83 billion yuan, 8.8% up from 2013. The marine chemical industry developed rapidly, realizing a full-year added value of 92.0 billion yuan, 15.4% up from 2013. The marine biomedicine industry continued to develop steadily, effecting an added value of 25.81 billion yuan, growing by 7.3% as compared with that in 2013. With the orderly advance of the offshore wind power generation projects, the marine electric power industry developed rapidly, affecting a full-year added value of 10.79 billion yuan, 17.6% up from 2013. As the level of industrialization of the seawater utilization industry was constantly raised, the industry continued to keep a posture of development, realizing a full-year added value of 1.27 billion yuan, 3.7% up from 2013. Speeding up the step of adjustment and transformation, the marine shipbuilding industry developed rapidly, bringing about a full-year added value of 139.55 billion yuan, growing by 15.5% as against that in 2013. The marine engineering construction industry continued to keep a steady growth, 7.5% up from 2013.

Tertiary marine industry

In 2014, the production situation of the coastal ports throughout the country is good on the whole, but owing to the accelerated imbalance between supply and demand, the domestic shipping market hovered between recession and recovery, and the marine communications and transport effected a value added of 533.7 billion yuan for the whole year, 9.6% up from 2013. The cargo handling capacity of coastal harbours was 803.307 million tons, registering a growth rate of 6.2% over 2013, and the handling capacity of international standardized containers, 7.1% up from 2013. The scale of coastal tourism continued to expand, while increasing input into the tourist infrastructure, the coastal regions constantly enriched their products of marine tourism, and marine passenger liners and yachting tourism stepped into a new phase of development.

The number of the inbound tourists received by the major coastal cities amounted to 38.866 million person-times, 1.070 million person-times more than that in 2013; through constantly tapping all potentialities of the domestic tourist market, the coastal tourism industry accomplished a full-year added value of 975.28 billion yuan, 11.1% up from 2013.

III. Development of Regional Marine Economy

In 2014, the three major economic zones – the Round-the-Bohai, Changjiang River Delta, and Zhujiang River Delta economic zones – quickened their step of structural adjustment and their marine economy continued to keep a posture of stable growth. The gross ocean products of the Round-the-Bohai Economic Zone, Changjiang River Delta Economic

Zone and Zhujiang River Delta Economic Zone were 2 228.89 billion yuan, 1 727.69 billion yuan and 1 322.98 billion yuan respectively, accounting for 36.7%, 28.5% and 21.8% of the national gross ocean product separately.

The overall marine economic growth rate of the Round-the-Bohai Economic Zone slowed down and its gross ocean product increased by 7.6% as compared with that in 2013 (at the current price), occupying 15.9% of the gross regional product. The value added of marine industries was 1 326.5 billion yuan and that of the marine-related industries 902.4 billion yuan. Marine fishery, offshore oil and gas industry, marine communications and transport, and coastal tourism were still the pillar industries marine economy in the zone, their added values adding up to 877.37 billion yuan and accounting for 84.8% of the regions major marine industries. The coastal tourism and the marine electric power industry effected a rapid growth, 23.4% and 17.4% up from 2013 respectively (at the current price).

The Changjiang River Delta Economic Zone presented a posture of growth and its gross ocean product increased by 10.9% as against that in 2013 (at the current price), accounting for 13.4% of the gross regional product. The value added of the marine industries in the zone was 1 024.38 billion yuan and that of the marine-related industries 703.31 billion yuan. In the light of the output value contributions, the industries of coastal tourism, marine communications and transport, marine shipbuilding and marine fishery ranked among the first, and the sum total of their added values accounted for 93.4% of the added value of the major marine industries, among which, the value added of coastal tourism occupied 42.6% of that of major marine industries and its output value contribution ranked first. The growth rates of the industries of marine shipbuilding, marine electric power, and marine engineering construction were faster, 23.0%, 23.0% and 19.5% up from 2013 separately (at the current price).

Marine economy in the Zhujiang River Delta Economic Zone continued to keep a steady growth, and its gross ocean product amounted to 1 322.98 billion yuan, 13.1% up from 2013 (at the current price) and accounting for 19.5% of the gross regional product. The value added of marine industries was 816.76 billion yuan and that of marine-related industries 506.21 yuan. Coastal tourism, marine communications and transport industry, marine chemical industry, and offshore oil and gas industry were still the pillar industries in the marine economic development of the zone, and the sum total of their added values accounting for 80.0% of the value added of major marine industries in this region.

IV. Marine Scientific Research and Education

In 2014, the marine scientific research and education continued to develop steadily. The statistical number of marine scientific research institutions totaled 177 with 37 679 employees, increasing by 4.6% as compared with 2013; the number of marine scientific and technological projects under-taken by these institutions was 17 702, 8.4% up from 2013; 16 908 marine scientific and technological papers and 314 kinds of marine scientific and technological works were published; the number of patents authorized was 4 020, of which, 1 883 were the ones for discovery, 17.9% up from 2013. The number of the institutions of higher learning which offer marine specialties amounted to 403, 3.8% up from last year, and the number of full-time teachers in this regard was 306 669, 2.5% up from the previous year. The number of graduates from the marine specialty of higher education and secondary vocational education was 82 631, 12.6% down from last year, and the number of students enrolled and that in school were 71 755 and 233 059 respectively, 3.8% and 9.0% down from the previous year separately.

V. Marine Environmental Protection and Disaster Mitigation

In 2014, the state of China's marine environmental quality was basically stable and remained the same level compared with 2013. Both the radioactivity level of seawater, the rate of air absorbed dose of the marine atmospheric γ radiation were normal. The ecological condition of the national marine protected areas (MPAs) basically kept stable. The environmental state in the key bathing beaches and coastal tourist resorts was good on the whole. The condition of environmental quality in the seawater culture and stock enhancement zone could basically satisfy the demand for cultivating activities. The environmental condition of the dumping zone was stable on the whole and had not been significantly affected by dumping activities, and the average eutrophication condition of nearshore waters was light. However, the environmental problems in the nearshore sea area such as water pollution, degradation of marine habitats, frequent occurrence of environmental disasters, etc. were still prominent. The pressure of terrigenous sewage discharge was great and the pollution of local nearshore sea area was still serious; 81.0% of the typical marine ecosystems that had been monitored such as estuaries, bays, etc. was in a sub-healthy or unhealthy state; the number of red tides discovered and their total area basically matched the average value in the recent five years; in the coastal plain area of the Bohai Sea, seawater intrusion and soil salinization were serious; and the degree of erosion in local sandy coastal sections and silty and muddy coasts was increased. In 2014, China's marine disasters were mainly storm surge, sea wave, sea ice, and red tide processes and other disasters such as green tide, coastal erosion, seawater intrusion and soil salinization, and salt-water tide also occurred in varying degrees. The direct economic loss caused by various marine disasters reached 13.614 billion yuan, which resulted in the death (missing) of 24 persons.

VI. Marine Administration

In 2014, marine administration was soundly carried forward and various marine efforts had obtained obvious results. A total of 5 011 certificates for the sea area use right were issued for the whole year, 27.3% up from 2013; the sea area with the ownership of patent rights reached 374 148 hm^2, 5.4% up from 2013; a total of 375 permits for oceanic dumping of dredged material were signed and issued throughout the year, and marine inspections for law inforcement were carried out on 145 170 occasions, during which 1 832 cases of unlawful practice were discovered; the marine numerical forecast service was provided for 29 395 times, 832 marine survey projects were carried out, acquiring 1.619 million data and the data quantity from various marine observations totaled 322.535 million and the quantity of satellite remote-sensing data received and placed on file for the whole year was 41 719.7 GB; the paper archives received this year totaled 21 640 volumes (copies), and the electronic archives 3 170 GB; and a total of 18 items of national standard and 94 items of professional standard had been authorized and examined, and 17 items of national standard and 12 items of professional standard issued.

图 1　全国海洋生产总值及三次产业构成
China's Gross Ocean Product and Three Industries Composition

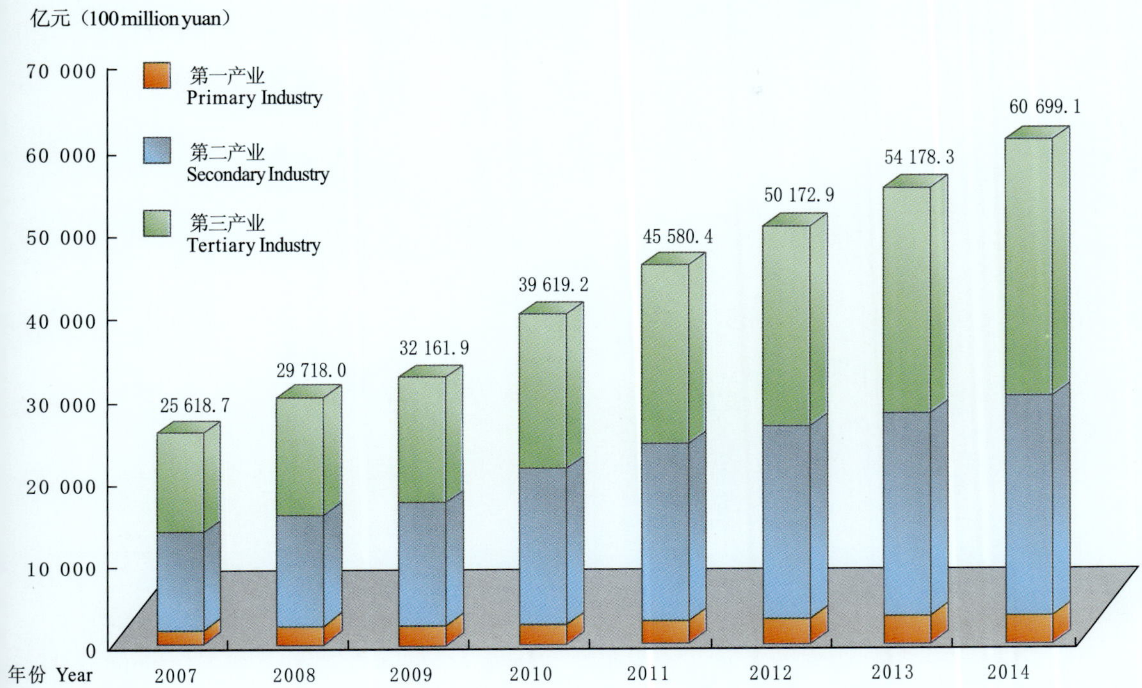

图 2　2014年全国主要海洋产业增加值构成
Composition of Added Values of the Major Marine Industries in China in 2014

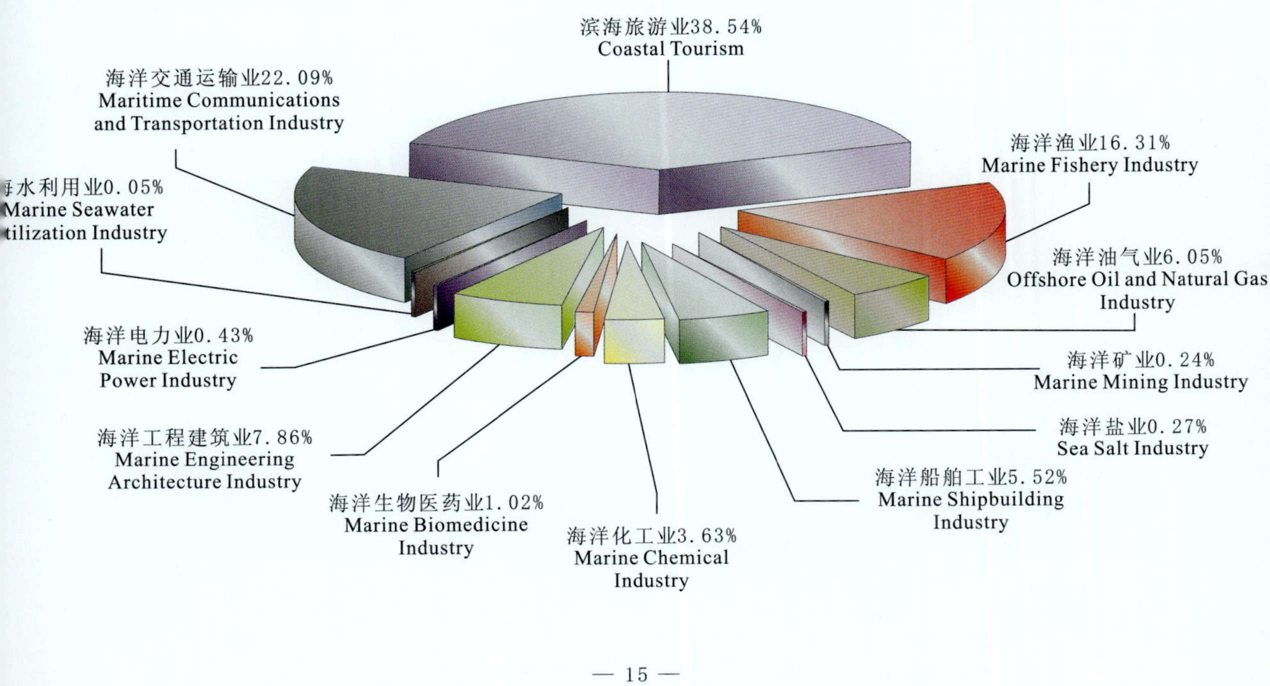

图 3 2014年沿海地区海洋生产总值
Gross Ocean Product by Coastal Regions in 2014

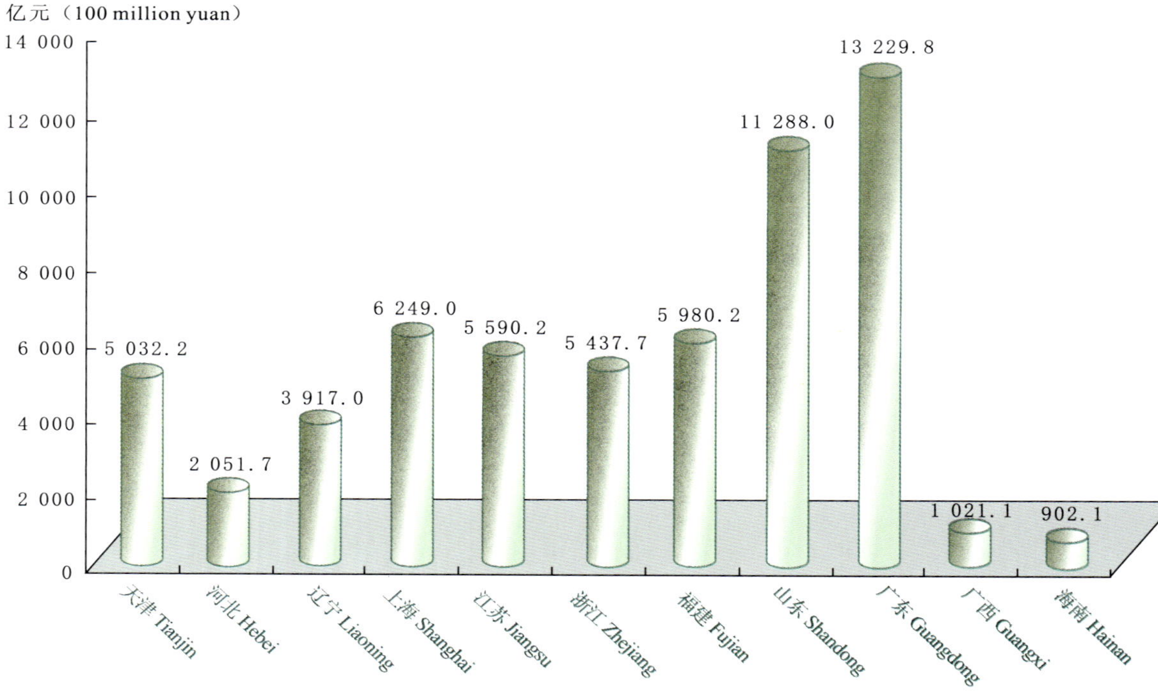

图 4 全国海洋捕捞和海水养殖产量
National Marine Catches and Mariculture Production

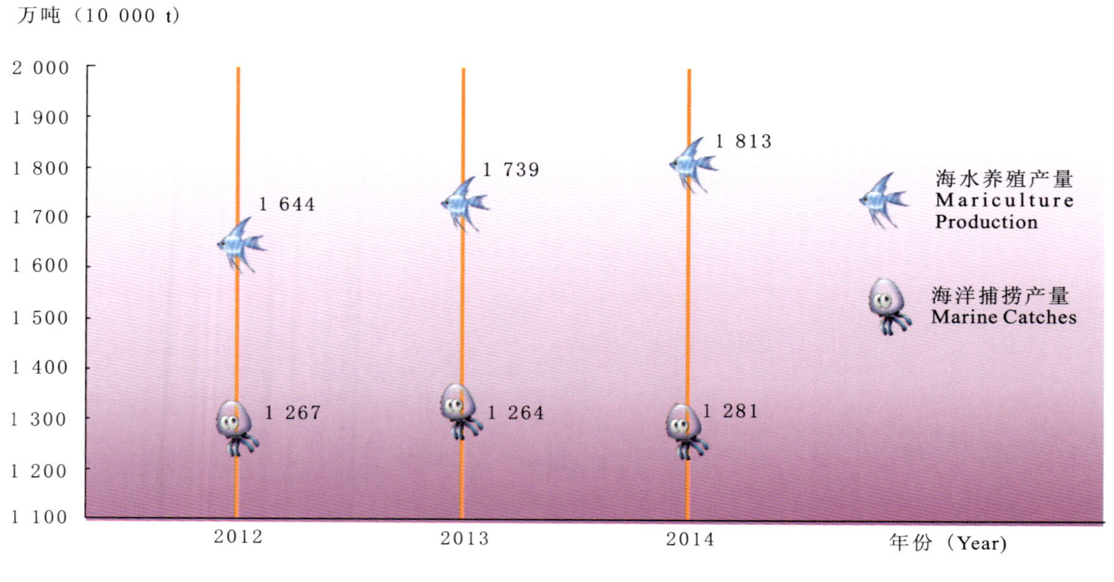

图 5 2014年沿海地区海洋捕捞和海水养殖产量
Marine Catches and Mariculture Production by Coastal Regions in 2014

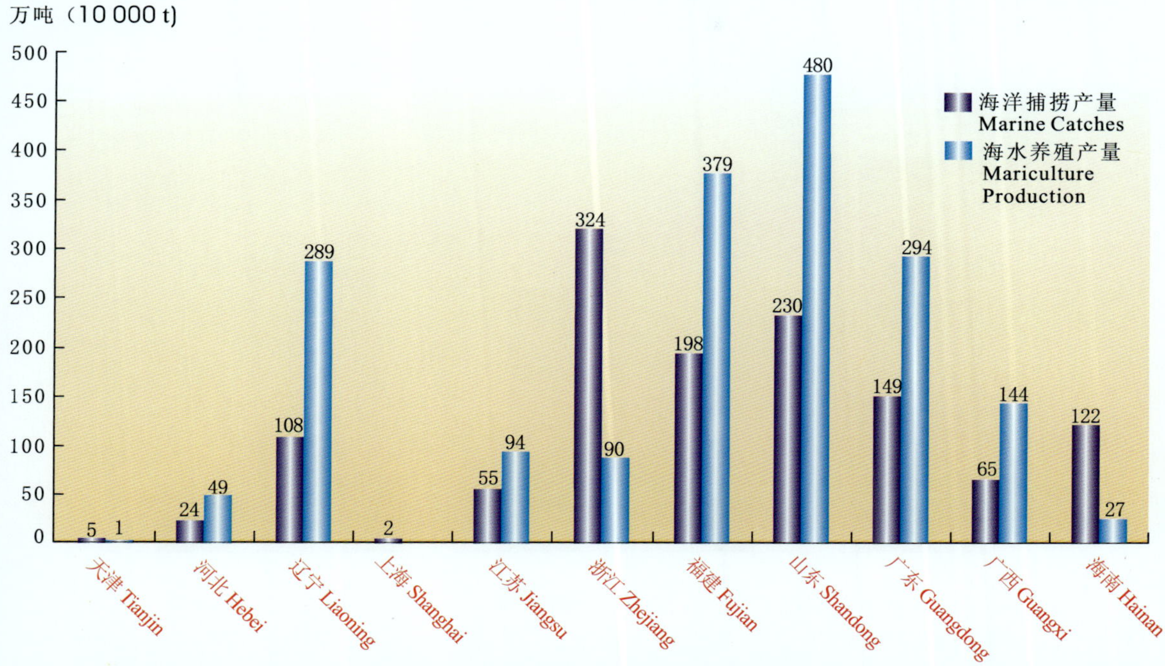

图 6 全国海洋石油和天然气产量
National Output of Offshore Oil and Natural Gas

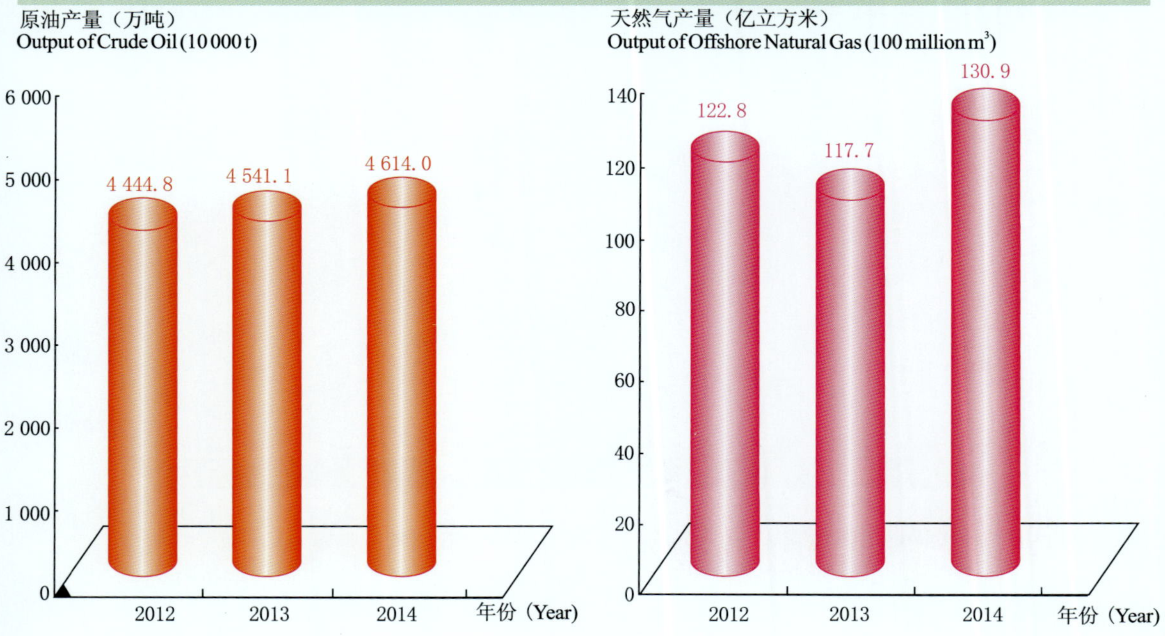

图 7 2014年海洋矿业产量

Production of Marine Mining Industry in 2014

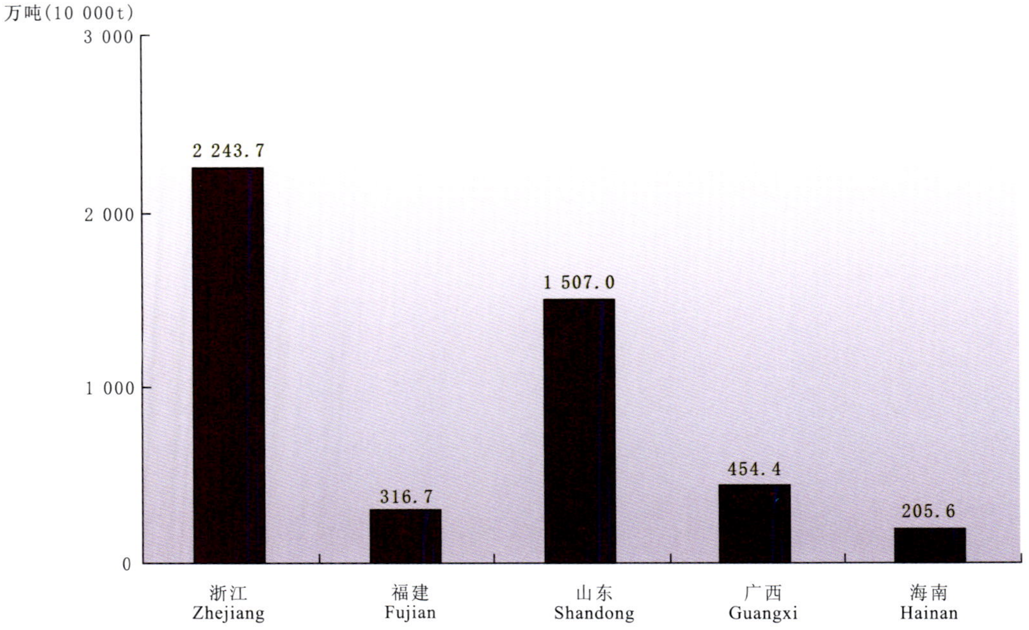

图 8 2014年沿海地区海盐产量

Sea Salt Production by Coastal Regions in 2014

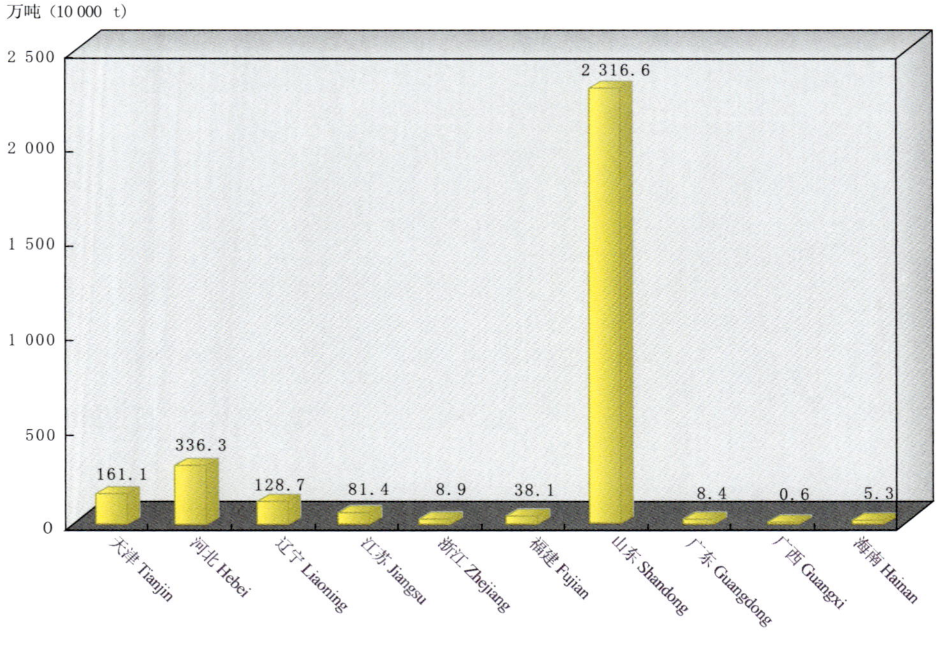

图 9　2014年沿海地区海洋造船完工量
Completed Quantity of Marine Shipbuilding in the Coastal Region in 2014

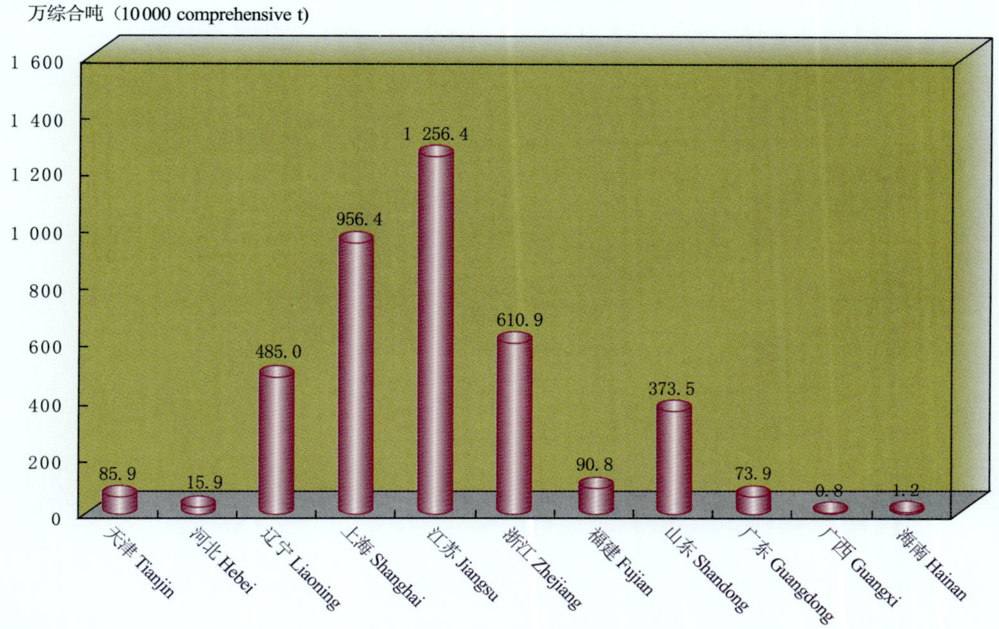

图 10　主要沿海城市接待入境旅游者人数
Number of Oversea Visitor Arrivals in Major Coastal Cities

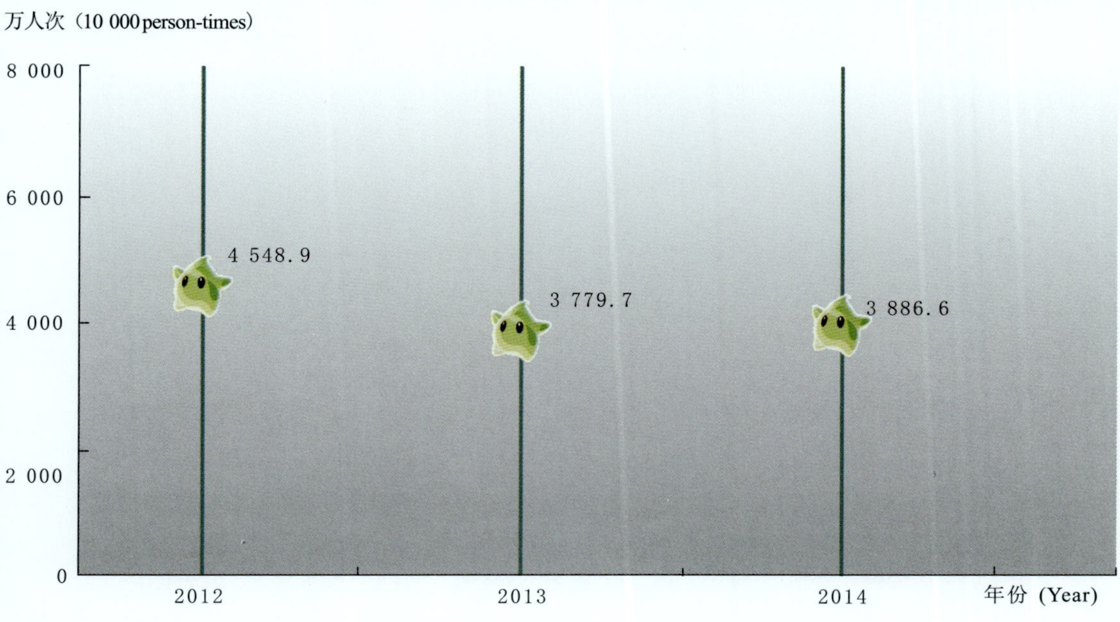

图 11 2014年沿海地区海洋货物周转量
Maritime Goods Turnover Volume by Coastal Regions in 2014

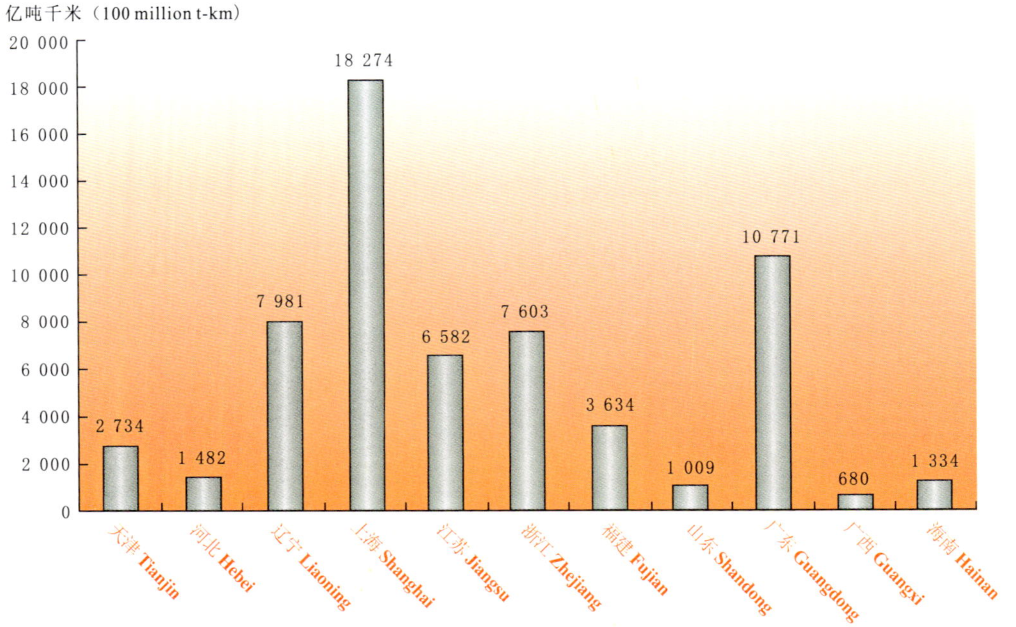

图 12 2014年沿海港口国际标准集装箱吞吐量
International Standardized Containers Handled by Coastal Seaports in 2014

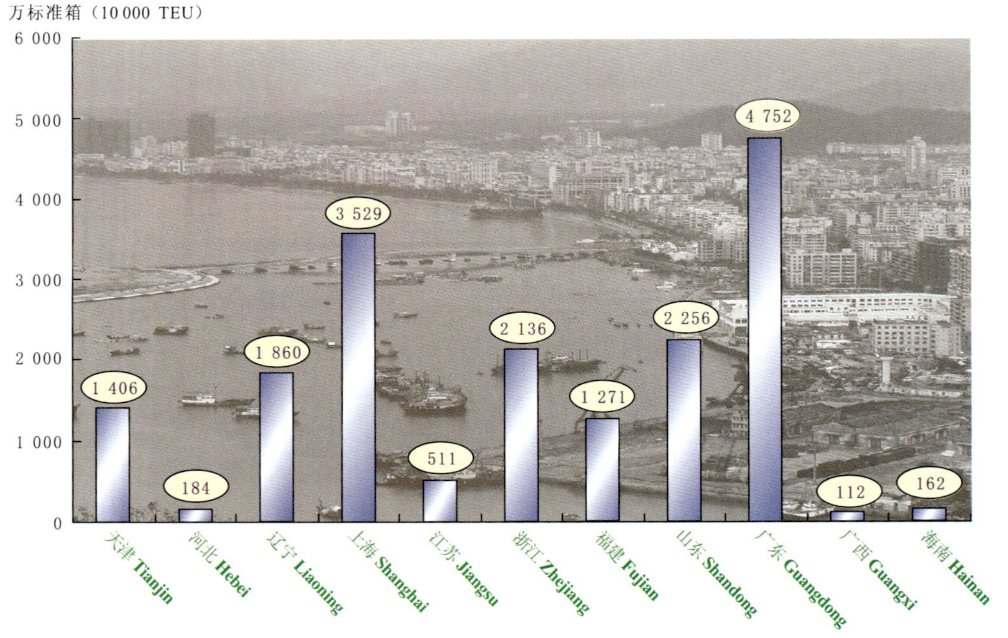

图 13　2014年三大经济区地区生产总值与海洋生产总值
GDP and GOP in the Three Major Economic Zones in 2014

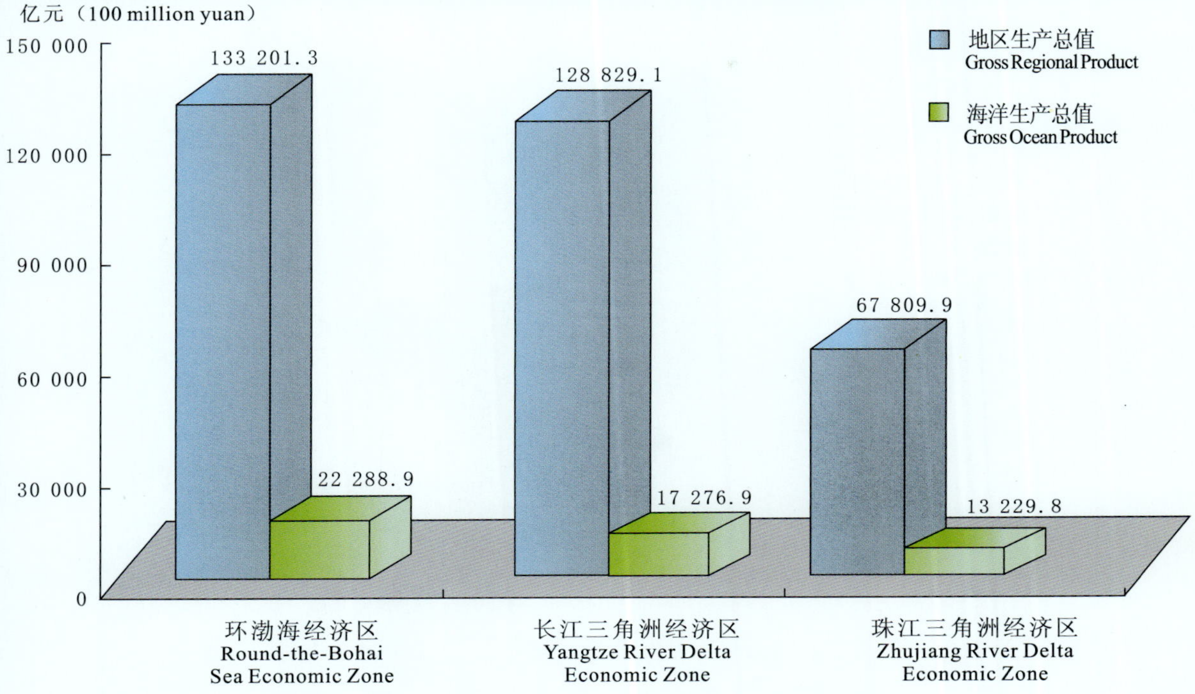

图 14　2014年三大经济区海洋生产总值占全国海洋生产总值比重
Proportion of the GOP of the Three Major Economic Zones in the National GOP in 2014

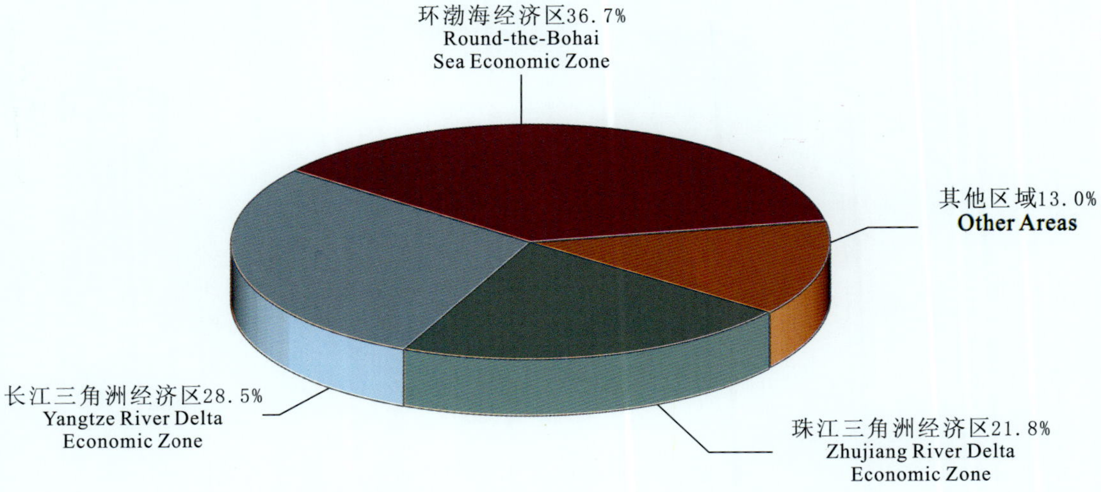

图 15 2014年沿海地区海洋生产总值

Gross Ocean Product by Coastal Regions in 2014

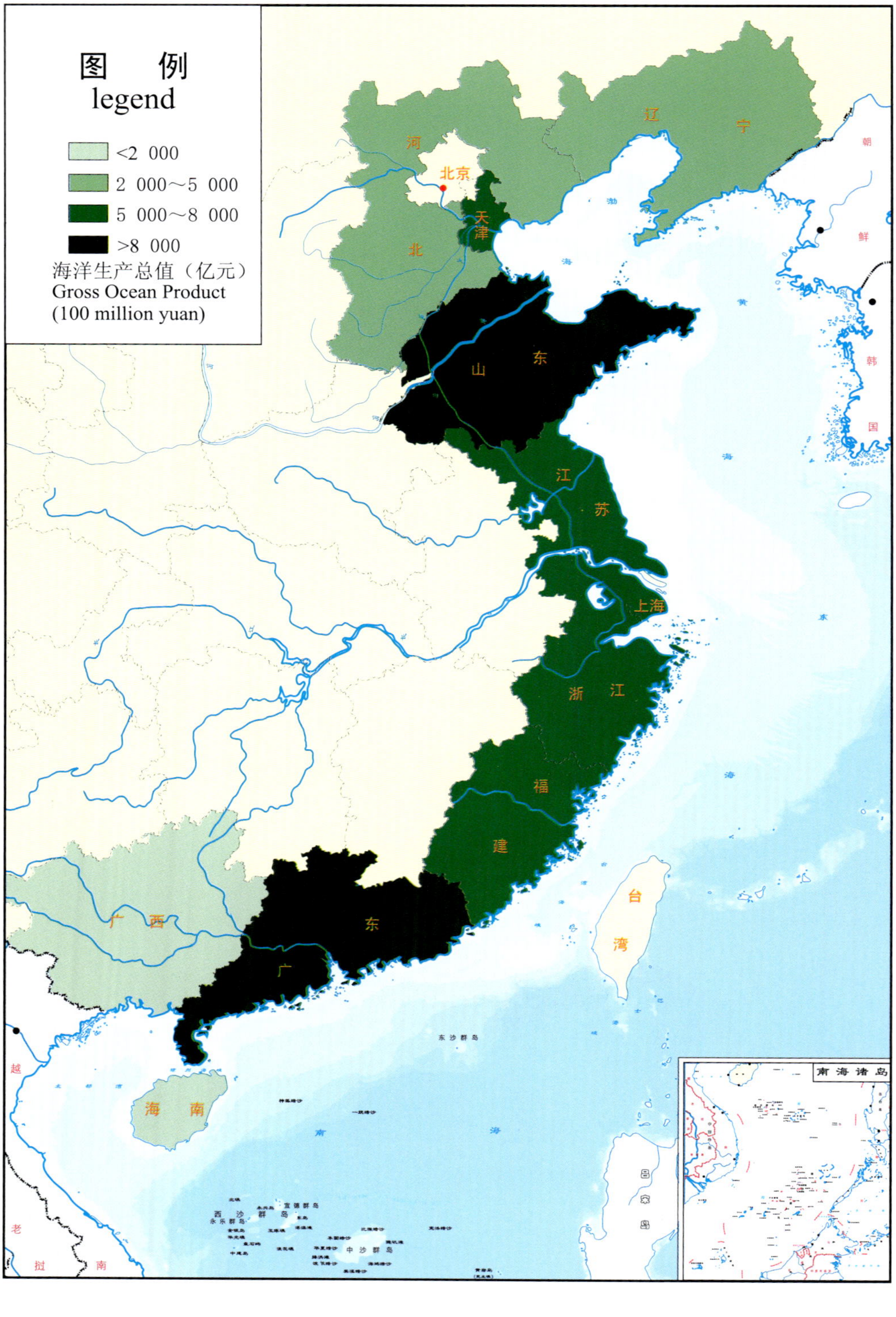

图 16 2014年沿海地区海洋经济贡献

Marine Economic Contributions by Coastal Regions in 2014

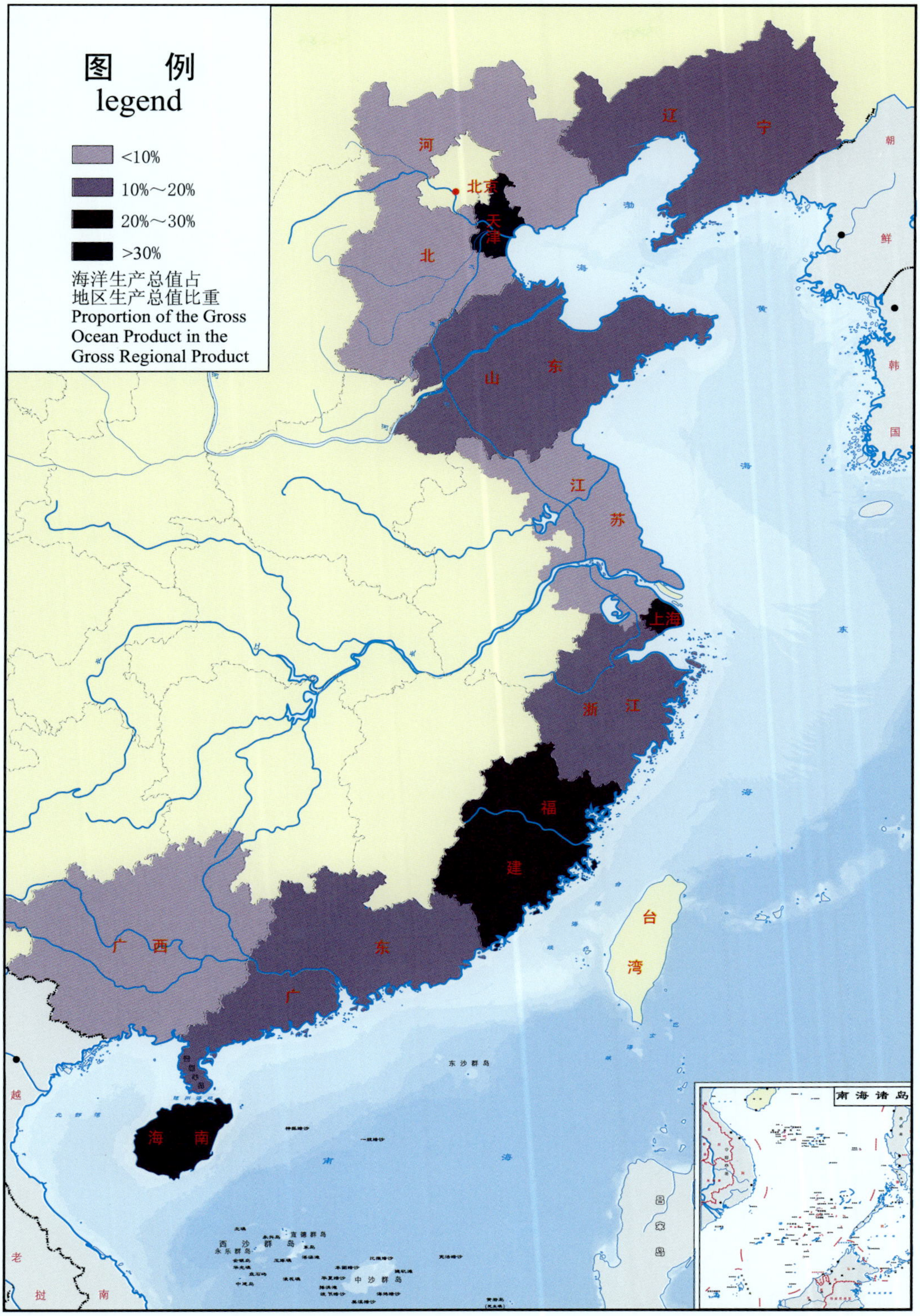

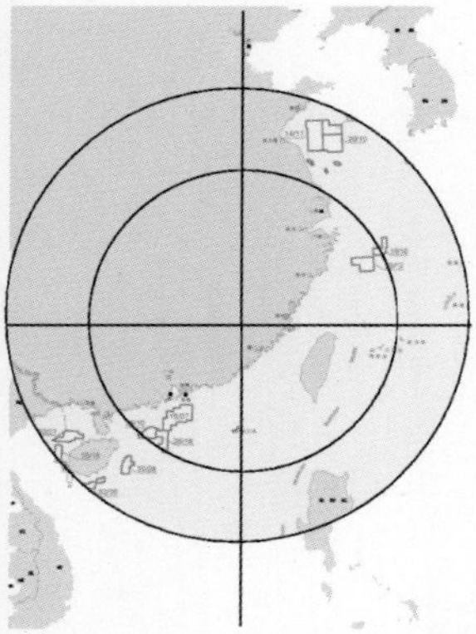

1

综 合 资 料

Integrated Data

1-1 沿海地区行政区划
Administrative Division of Coastal Regions

单位：个 (number)

沿海地区 Coastal Region	沿海城市 Coastal City	沿海地带 Coastal County (District)			
		合计 Total	县 County	县级市 County-level City	区 District
合计 Total	**54**	**238**	**59**	**53**	**126**
天津 Tianjin	1	1			1
河北 Hebei	3	11	6	1	4
辽宁 Liaoning	6	22	4	7	11
上海 Shanghai	1	5	1		4
江苏 Jiangsu	3	15	7	4	4
浙江 Zhejiang	7	35	10	9	16
福建 Fujian	6	34	11	8	15
山东 Shandong	7	35	5	12	18
广东 Guangdong	14	55	9	6	40
广西 Guangxi	3	8	1	1	6
海南 Hainan	3	17	5	5	7

注：沿海地带中未包括广东省的东莞、中山。

Note: The coastal zone does not include the regions under the jurisdiction of Dongguan city and Zhongshan city.

1-2 沿海行政区划一览表
Table of Administrative Division of Coastal Regions

沿海地区 Coastal Region	地区代码 Zip Code	沿海城市 Coastal City	地区代码 Zip Code	沿海地带 Coastal County (District)	地区代码 Zip Code
天 津 Tianjin	120000			滨海新区Binhai Xinqu	120116
河 北 Hebei	130000	唐山 Tangshan	130200	丰南区Fengnan Qu	130207
				曹妃甸区Caofeidian Qu	130209
				滦南县Luannan Xian	130224
				乐亭县Leting Xian	130225
		秦皇岛 Qinhuangdao	130300	海港区Haigang Qu	130302
				山海关区Shanhaiguan Qu	130303
				北戴河区Beidaihe Qu	130304
				昌黎县Changli Xian	130322
				抚宁县Funing Xian	130323
		沧州 Cangzhou	130900	海兴县Haixing Xian	130924
				黄骅市Huanghua Shi	130983
辽 宁 Liaoning	210000	大连 Dalian	210200	中山区Zhongshan Qu	210202
				西岗区Xigang Qu	210203
				沙河口区Shahekou Qu	210204
				甘井子区Ganjingzi Qu	210211
				旅顺口区Lüshunkou Qu	210212
				金州区Jinzhou Qu	210213
				长海县Changhai Xian	210224
				瓦房店市Wafangdian Shi	210281
				普兰店市Pulandian Shi	210282
				庄河市Zhuanghe Shi	210283
		丹东 Dandong	210600	振兴区Zhenxing Qu	210603
				东港市Donggang Shi	210681
		锦州 Jinzhou	210700	凌海市Linghai Shi	210781
		营口 Yingkou	210800	鲅鱼圈区Bayuquan Qu	210804
				老边区Laobian Qu	210811
				盖州市Gaizhou Shi	210881
		盘锦 Panjin	211100	大洼县Dawa Xian	211121
				盘山县Panshan Xian	211122

1-2 续表1 continued

沿海地区 Coastal Region	地区代码 Zip Code	沿海城市 Coastal City	地区代码 Zip Code	沿海地带 Coastal County (District)	地区代码 Zip Code
		葫芦岛 Huludao	211400	连山区Lianshan Qu	211402
				龙港区Longgang Qu	211403
				绥中县Suizhong Xian	211421
				兴城市Xingcheng Shi	211481
上海 Shanghai	310000			宝山区Baoshan Qu	310113
				浦东新区Pudong Xinqu	310115
				金山区Jinshan Qu	310116
				奉贤区Fengxian Qu	310120
				崇明县Chongming Xian	310230
江苏 Jiangsu	320000	南通 Nantong	320600	通州区Tongzhou Qu	320612
				海安县Hai'an Xian	320621
				如东县Rudong Xian	320623
				启东市Qidong Shi	320681
				海门市Haimen Shi	320684
		连云港 Lianyungang	320700	连云区Lianyun Qu	320703
				海州区Haizhou Qu	320706
				赣榆区Ganyu Qu	320707
				灌云县Guanyun Xian	320723
				灌南县Guannan Xian	320724
		盐城 Yancheng	320900	响水县Xiangshui Xian	320921
				滨海县Binhai Xian	320922
				射阳县Sheyang Xian	320924
				东台市Dongtai Shi	320981
				大丰市Dafeng Shi	320982
浙江 Zhejiang	330000	杭州 Hangzhou	330100	滨江区Binjiang Qu	330108
				萧山区Xiaoshan Qu	330109
		宁波 Ningbo	330200	海曙区Haishu Qu	330203
				江东区Jiangdong Qu	330204
				江北区Jiangbei Qu	330205
				北仑区Beilun Qu	330206
				镇海区Zhenhai Qu	330211
				鄞州区Yinzhou Qu	330212
				象山县Xiangshan Xian	330225
				宁海县Ninghai Xian	330226
				余姚市Yuyao Shi	330281
				慈溪市Cixi Shi	330282
				奉化市Fenghua Shi	330283

1-2 续表2 continued

沿海地区 Coastal Region	地区代码 Zip Code	沿海城市 Coastal City	地区代码 Zip Code	沿海地带 Coastal County (District)	地区代码 Zip Code
		温州 Wenzhou	330300	龙湾区Longwan Qu	330303
				瓯海区Ouhai Qu	330304
				洞头县Dongtou Xian	330322
				平阳县Pingyang Xian	330326
				苍南县Cangnan Xian	330327
				瑞安市Rui'an Shi	330381
				乐清市Yueqing Shi	330382
		嘉兴 Jiaxing	330400	海盐县Haiyan Xian	330424
				海宁市Haining Shi	330481
				平湖市Pinghu Shi	330482
		绍兴 Shaoxing	330600	柯桥区Keqiao Qu	330603
				上虞区Shangyu Qu	330604
		舟山 Zhoushan	330900	定海区Dinghai Qu	330902
				普陀区Putuo Qu	330903
				岱山县Daishan Xian	330921
				嵊泗县Shengsi Xian	330922
		台州 Taizhou	331000	椒江区Jiaojiang Qu	331002
				路桥区Luqiao Qu	331004
				玉环县Yuhuan Xian	331021
				三门县Sanmen Xian	331022
				温岭市Wenling Shi	331081
				临海市Linhai Shi	331082
福建 Fujian	350000	福州 Fuzhou	350100	马尾区Mawei Qu	350105
				连江县Lianjiang Xian	350122
				罗源县Luoyuan Xian	350123
				平潭县Pingtan Xian	350128
				福清市Fuqing Shi	350181
				长乐市Changle Shi	350182
		厦门 Xiamen	350200	思明区Siming Qu	350203
				海沧区Haicang Qu	350205

1-2 续表3 continued

沿海地区 Coastal Region	地区代码 Zip Code	沿海城市 Coastal City	地区代码 Zip Code	沿海地带 Coastal County (District)	地区代码 Zip Code
				湖里区Huli Qu	350206
				集美区Jimei Qu	350211
				同安区Tong'an Qu	350212
				翔安区Xiang'an Qu	350213
		莆田 Putian	350300	城厢区Chengxiang Qu	350302
				涵江区Hanjiang Qu	350303
				荔城区Licheng Qu	350304
				秀屿区Xiuyu Qu	350305
				仙游县Xianyou Xian	350322
		泉州 Quanzhou	350500	丰泽区Fengze Qu	350503
				洛江区Luojiang Qu	350504
				泉港区Quangang Qu	350505
				惠安县Hui'an Xian	350521
				金门县Jinmen Xian	350527
				石狮市Shishi Shi	350581
				晋江市Jinjiang Shi	350582
				南安市Nan'an Shi	350583
		漳州 Zhangzhou	350600	云霄县Yunxiao Xian	350622
				漳浦县Zhangpu Xian	350623
				诏安县Zhao'an Xian	350624
				东山县Dongshan Xian	350626
				龙海市Longhai Shi	350681
		宁德 Ningde	350900	蕉城区Jiaocheng Qu	350902
				霞浦县Xiapu Xian	350921
				福安市Fu'an Shi	350981
				福鼎市Fuding Shi	350982
山 东 Shandong	370000	青岛 Qingdao	370200	市南区Shinan Qu	370202
				市北区Shibei Qu	370203
				黄岛区Huangdao Qu	370211
				崂山区Laoshan Qu	370212
				李沧区Licang Qu	370213
				城阳区Chengyang Qu	370214
				胶州市Jiaozhou Shi	370281
				即墨市Jimo Shi	370282

1-2 续表4 continued

沿海地区 Coastal Region	地区代码 Zip Code	沿海城市 Coastal City	地区代码 Zip Code	沿海地带 Coastal County (District)	地区代码 Zip Code
		东营 Dongying	370500	东营区Dongying Qu	370502
				河口区Hekou Qu	370503
				垦利县Kenli Xian	370521
				利津县Lijin Xian	370522
				广饶县Guangrao Xian	370523
		烟台 Yantai	370600	芝罘区Zhifu Qu	370602
				福山区Fushan Qu	370611
				牟平区Muping Qu	370612
				莱山区Laishan Qu	370613
				长岛县Changdao Xian	370634
				龙口市Longkou Shi	370681
				莱阳市Laiyang Shi	370682
				莱州市Laizhou Shi	370683
				蓬莱市Penglai Shi	370684
				招远市Zhaoyuan Shi	370685
				海阳市Haiyang Shi	370687
		潍坊 Weifang	370700	寒亭区Hanting Qu	370703
				寿光市Shouguang Shi	370783
				昌邑市Changyi Shi	370786
		威海 Weihai	371000	环翠区Huancui Qu	371002
				文登区Wendeng Qu	371003
				荣成市Rongcheng Shi	371082
				乳山市Rushan Shi	371083
		日照 Rizhao	371100	东港区Donggang Qu	371102
				岚山区Lanshan Qu	371103
		滨州 Binzhou	371600	沾化区Zhanhua Qu	371603
				无棣县Wudi Xian	371623
广东 Guangdong	440000	广州 Guangzhou	440100	荔湾区Liwan Qu	440103
				越秀区Yuexiu Qu	440104
				海珠区Haizhu Qu	440105
				天河区Tianhe Qu	440106
				白云区Baiyun Qu	440111
				黄埔区Huangpu Qu	440112
				番禺区Panyu Qu	440113
				南沙区Nansha Qu	440115
				萝岗区Luogang Qu	440116

1-2 续表5 continued

沿海地区 Coastal Region	地区代码 Zip Code	沿海城市 Coastal City	地区代码 Zip Code	沿海地带 Coastal County (District)	地区代码 Zip Code
		深圳 Shenzhen	440300	罗湖区Luohu Qu	440303
				福田区Futian Qu	440304
				南山区Nanshan Qu	440305
				宝安区Bao'an Qu	440306
				龙岗区Longgang Qu	440307
				盐田区Yantian Qu	440308
		珠海 Zhuhai	440400	香洲区Xiangzhou Qu	440402
				斗门区Doumen Qu	440403
				金湾区Jinwan Qu	440404
		汕头 Shantou	440500	龙湖区Longhu Qu	440507
				金平区Jinping Qu	440511
				濠江区Haojiang Qu	440512
				潮阳区Chaoyang Qu	440513
				潮南区Chaonan Qu	440514
				澄海区Chenghai Qu	440515
				南澳县Nan'ao Xian	440523
		江门 Jiangmen	440700	蓬江区Pengjiang Qu	440703
				江海区Jianghai Qu	440704
				新会区Xinhui Qu	440705
				台山市Taishan Shi	440781
				恩平市Enping Shi	440785
		湛江 Zhanjiang	440800	赤坎区Chikan Qu	440802
				霞山区Xiashan Qu	440803
				坡头区Potou Qu	440804
				麻章区Mazhang Qu	440811
				遂溪县Suixi Xian	440823
				徐闻县Xuwen Xian	440825
				廉江市Lianjiang Shi	440881
				雷州市Leizhou Shi	440882
				吴川市Wuchuan Shi	440883
		茂名 Maoming	440900	茂南区Maonan Qu	440902
				电白区Dianbai Qu	440904
		惠州 Huizhou	441300	惠城区Huicheng Qu	441302
				惠阳区Huiyang Qu	441303
				惠东县Huidong Xian	441323
		汕尾 Shanwei	441500	城　区Chengqu	441502
				海丰县Haifeng Xian	441521
				陆丰市Lufeng Shi	441581

1-2 续表6 continued

沿海地区 Coastal Region	地区代码 Zip Code	沿海城市 Coastal City	地区代码 Zip Code	沿海地带 Coastal County (District)	地区代码 Zip Code
		阳江 Yangjiang	441700	江城区Jiangcheng Qu	441702
				阳西县Yangxi Xian	441721
				阳东县Yangdong Xian	441723
		东莞 Dongguan	441900		
		中山 Zhongshan	442000		
		潮州 Chaozhou	445100	湘桥区Xiangqiao Qu	445102
				饶平县Raoping Xian	445122
		揭阳 Jieyang	445200	榕城区Rongcheng Qu	445202
				揭东区Jiedong Qu	445203
				惠来县Huilai Xian	445224
广 西 Guangxi	450000	北海 Beihai	450500	海城区Haicheng Qu	450502
				银海区Yinhai Qu	450503
				铁山港区Tieshangang Qu	450512
				合浦县Hepu Xian	450521
		防城港 Fangchenggang	450600	港口区Gangkou Qu	450602
				防城区Fangcheng Qu	450603
				东兴市Dongxing Shi	450681
		钦州 Qinzhou	450700	钦南区Qinnan Qu	450702
海 南 Hainan	460000	海口 Haikou	460100	秀英区Xiuying Qu	460105
				龙华区Longhua Qu	460106
				美兰区Meilan Qu	460108
		三亚 Sanya	460200	海棠区Haitang Qu	460207
				吉阳区Jiyang Qu	460203
				天涯区Tianya Qu	460204
				崖州区YaZhou Qu	460205
		省直辖县 Counties Directly under the Hainan Province Government	469000	琼海市Qionghai Shi	469002
				儋州市Danzhou Shi	469003
				文昌市Wenchang Shi	469005
				万宁市Wanning Shi	469006
				东方市Dongfang Shi	469007
				澄迈县Chengmai Xian	469023
				临高县Lingao Xian	469024
				昌江黎族自治县Changjiang Lizu Zizhixian	469026
				乐东黎族自治县 Ledong Lizu Zizhixian	469027
				陵水黎族自治县 Lingshui Lizu Zizhixian	469028

1-3 海洋自然地理
Marine Physical Geography

指　　标	Item	指标值 Data
海洋平均深度　（米）	Average Depth of Sea (m)	961
海洋最大深度　（米）	Maximum Depth of Sea (m)	5 559
岸线总长度　（千米）	Total Length of Coastline (km)	32 000
大陆岸线长度	Length of Continental Coastline	18 000
岛屿岸线长度	Length of Insular Coastline	14 000
＞500m^{2}岛屿数　（个）	Number of Islands ＞500 m^2 each (unit)	7 300
岛屿面积　（万平方千米）	Area of Islands (10 000 km^2)	8
已利用的无居民海岛　（个）	Uninhabited Islands which Have Been Utilized (unit)	1 900
特殊用途海岛	Islands Used for Special Purposes	1 020
公共服务用岛	Islands Used for Public Service	365
旅游娱乐用岛	Islands Used for Tourism and Recreation	73
农林牧渔业用岛	Islands Used for Agriculture, Forestry, Animal Husbandry and Fishery	340
工业、仓储、交通运输用岛	Islands Used for Industry, Storage, Communications and Transport	49

注：海岛数据来源于《全国海岛保护规划》。

Note: Data on islands are derived from the *National Plan for Island Protection*.

1-4 海区海洋石油天然气储量
Offshore Oil and Gas Reserves in the Sea Area

海 区 Sea Area	海洋石油（万吨） Offshore Oil (10 000 t)		海洋天然气（亿立方米） Natural Gas(100 million m^3)	
	累计探明技术可采储量 Proven Technically Recoverable Reserves in the Aggregate	剩余技术可采储量 Surplus Technically Recoverable Reserves	累计探明技术可采储量 Proven Technically Recoverable Reserves in the Aggregate	剩余技术可采储量 Surplus Technically Recoverable Reserves
合 计 Total	**120 958.1**	**55 798.2**	**5 426.3**	**4 099.3**
渤 海 Bohai Sea	73 513.9	42 524.7	739.9	491.6
黄 海 Yellow Sea				
东 海 East China Sea	1 641.8	990.1	1 196.9	1 098.0
南 海 South China Sea	45 802.4	12 283.4	3 489.5	2 509.8

注：数据来源于《2014年全国矿产资源储量通报》。

Note: The data come from the *Journal on the National Mineral Resources Reserves in 2014*.

1-5 沿海地区水资源情况
Water Resources by Coastal Regions

地 区 Region	水资源总量（亿立方米） Total Amount of Water Resources (100 million m^3)	地表水资源量 Surface Water Resources	地下水资源量 Groundwater Resources	地表水与地下水资源重复量 Duplicate Measurement of Surface Water and Groundwater	人均水资源量（立方米/人） Per Capita Water Resources (m^3/person)
全国总计 National Total	**27 266.9**	**26 263.9**	**7 745.0**	**6 742.0**	**1 998.6**
天 津 Tianjin	11.4	8.3	3.7	0.6	76.1
河 北 Hebei	106.2	46.9	89.3	30.1	144.3
辽 宁 Liaoning	145.9	123.7	82.3	60.1	332.4
上 海 Shanghai	47.1	40.1	10.0	3.0	194.8
江 苏 Jiangsu	399.3	296.4	118.9	16.0	502.3
浙 江 Zhejiang	1 132.1	1 118.2	231.8	217.9	2 057.3
福 建 Fujian	1 219.6	1 218.4	330.5	329.3	3 218.0
山 东 Shandong	148.4	76.6	116.9	45.0	152.1
广 东 Guangdong	1 718.4	1 709.0	420.5	411.1	1 608.4
广 西 Guangxi	1 990.9	1 989.6	403.0	401.7	4 203.3
海 南 Hainan	383.5	378.7	96.7	91.9	4 266.0

注： 数据来源于《2015中国统计年鉴》。

Note: The data come from the *China Statistical Yearbook 2015*.

1-6　沿海地区湿地面积
Area of Wetlands by Coastal Regions

地　区 Region	湿地总面积 （千公顷） Total Area of Wetlands (1 000 hm^2)	#近海与海岸 Coasts and Seashores
合　计 Total	**12 466.0**	**5 795.9**
天 津 Tianjin	295.6	104.3
河 北 Hebei	941.9	231.9
辽 宁 Liaoning	1 394.8	713.2
上 海 Shanghai	464.6	386.6
江 苏 Jiangsu	2 822.8	1 087.5
浙 江 Zhejiang	1 110.1	692.5
福　建 Fujian	871.0	575.6
山　东 Shandong	1 737.5	728.5
广　东 Guangdong	1 753.4	815.1
广　西 Guangxi	754.3	259.0
海　南 Hainan	320.0	201.7

注：数据来源于《2015中国统计年鉴》。

Note: The data come from the *China Statistical Yearbook 2015*.

1-7 红树林各地类面积

Site Classification and Area of Sharpleaf Mangrove (*Rhizophora Apiculata*)

单位：公顷 (hm²)

地 区 Region	红树林各地类总面积 Total Site Area of Sharpleaf Mangrove	现有面积 Established	未成林面积 Unestablished	宜林地面积 Suitable for Planting
全国总计 National Total	**82 757.2**	**22 024.9**	**1 884.1**	**58 848.2**
浙 江 Zhejiang	5 452.3	20.6	236.1	5 195.6
福 建 Fujian	13 410.1	615.1	286.4	12 508.6
广 东 Guangdong	32 325.9	9 084.0	981.3	22 260.6
广 西 Guangxi	18 029.2	8 374.9	380.3	9 274.0
海 南 Hainan	13 539.7	3 930.3		9 609.4

注： 数据来源于国家林业局专项调查。

Note: The data come from Special Survey of State Forestry Administration.

1-8 主要沿海城市气候基本情况
Climate of Major Coastal Cities

城市 City	年平均气温（摄氏度） Annual Average Temperature(℃)	年平均相对湿度（%） Annual Average Relative Humidity (%)	全年降水量（毫米） Annual Precipitation (mm)	全年日照时数（小时） Annual Sunshine Hours (h)
天津 Tianjin	14.0	59	441.4	2 265.6
上海 Shanghai	17.0	73	1 295.3	1 612.6
杭州 Hangzhou	17.5	73	1 359.9	1 407.2
福州 Fuzhou	20.8	72	1 628.0	1 591.7
广州 Guangzhou	21.7	79	2 234.0	1 613.6
海口 Haikou	24.7	83	1 861.3	2 112.0

注：数据来源于《2015中国统计年鉴》。

Note: The data come from the *China Statistical Yearbook 2015*.

主要统计指标解释

1. 沿海地区 即广义的沿海地区，是指有海岸线（大陆岸线和岛屿岸线）的地区，按行政区划分为沿海省、自治区、直辖市。

2. 沿海城市 是指有海岸线的直辖市和地级市（包括其下属的全部区、县和县级市）。

3. 沿海地带 即狭义的沿海地区，是指有海岸线的县、县级市、区（包括直辖市和地级市的区）。

4. 海洋 是海和洋的统称。洋为地球表面上相连接的广大咸水水体的主体部分。海为地球表面相连接的广大咸水水体被陆地、岛礁、半岛包围或分隔的边缘部分。

5. 水资源总量 指评价区内降水形成的地表和地下产水总量，即地表产流量与降水入渗补给地下水量之和，不包括过境水量。

6. 地表水资源量 指评价区内河流、湖泊、冰川等地表水体中可以逐年更新的动态水量，即当地天然河川径流量。

7. 地下水资源量 指评价区内降水和地表水对饱水岩土层的补给量，包括降水入渗补给量和河道、湖库、渠系、渠灌田间等地表水体的入渗补给量。

8. 地表水与地下水资源重复量 指地表水和地下水相互转化的部分，即天然河川径流量中的地下水排泄量和地下水补给量中来源于地表水的入渗补给量。

9. 湿地 指天然或人工、长久或暂时性的沼泽地、泥炭地或水域地带，包括静止或流动、淡水、半咸水、咸水体，低潮时水深不超过6米的水域以及海岸地带地区的珊瑚滩和海草床、滩涂、红树林、河口、河流、淡水沼泽、沼泽森林、湖泊、盐沼及盐湖。

10. 红树林 指生长在热带、亚热带低能海岸潮间带上部，受周期性潮水浸淹，以红树植物为主体的常绿灌木或乔木组成的潮滩湿地木本生物群落。

11. 气温 指空气的温度，我国一般以摄氏度(℃)为单位表示。气象观测的温度表是放在离地面约1.5米处通风良好的百叶箱里测量的，因此，通常说的气温指的是离地面1.5米处百叶箱中的温度。其统计计算方法为：

月平均气温是将全月各日的平均气温相加，除以该月的天数而得；

年平均气温是将12个月的月平均气温累加后除以12而得。

12. 相对湿度 指空气中实际所含水蒸气密度和同温度下饱和水蒸气密度的百分比值。其统计方法与气温相同。

13. 降水量 指从天空降落到地面的液态或固态(经融化后)水，未经蒸发、渗透、流失而在地面上积聚的深度。其统计计算方法为：

月降水量是将全月各日的降水量累加而得；

年降水量是将12个月的月降水量累加而得。

14. 日照时数 指太阳实际照射地面的时间。其统计方法与降水量相同。

Explanatory Notes on Main Statistical Indicators

1. Coastal Region, i.e., the coastal region in a broad sense, refers to the regions with coastlines (continental and island coastlines), which are divided into the coastal provinces, autonomous regions and municipalities directly under the Central Government according to the administrative zoning.

2. Coastal City refers to the municipalities directly under the Central Government and the prefecture-level cities (including all the districts, counties and county-level cities under them).

3. Coastal Zone, i.e., the coastal region in a narrow sense, refers to the counties, county-level cities and districts with coastlines (including the districts under the municipalities directly under the Central Government and the prefecture-level districts).

4. Ocean is the general name for sea and ocean. Ocean refers to the main body of large salt water connected with the earth surface. Sea refers to the edge areas of the salt water on the earth surface that are compartmentalized or surrounded by land, island, reef or peninsula.

5. Total Water Resources refers to total volume of water resources measured as run-off for surface water from rainfall and recharge for groundwater in a given area, excluding transit water.

6. Surface Water Resources refers to total renewable resources which exist in rivers, lakes, glaciers and other collectors from rainfall and are measured as run-off of rivers.

7. Groundwater Resources refers to replenishment of aquifers with rainfall and surface water.

8. Duplicated Measurement between Surface Water and Groundwater refers to the exchange between surface water and groundwater, i.e. run-off of rivers includes some depletion into groundwater while groundwater includes some replenishment from surface water.

9. Wetlands refer to marshland and peat bog, whether natural or man-made, permanent or temporary; water covered areas, whether stagnant or flowing, with fresh or brackish-fresh or salty water that is less than 6 meters deep at low tide; as well as coral beach, weed beach, mud beach, mangrove, river outlet, rivers, fresh-water marshland, marshland forests, lakes, salty bog and salt lakes along the coastal areas.

10. Mangrove refers to evergreen woody plants or plant communities in tropical or sub-tropical zones which live between the sea and the land in areas which are inundated by tides.

11. Temperature refers to the air temperature. China uses centigrade as the unit. The thermometry used for weather observation is put in a breezy shutter, which is 1.5 meters high from the ground. Therefore, the commonly used temperature refers to the temperature in the breezy shutter 1.5 meters away from the ground. The calculation method is as follows:

Monthly average temperature is the summation of average daily temperature of one month divided by the actual days of that particular month.

Annual average temperature is the summation of monthly averages of a year divided by 12 months.

12. Relative Humidity refers to the ratio of actual water vapour pressure to the saturated water vapour density under the current temperature. The calculation method is the same as that of temperature.

13. Volume of Precipitation refers to the deepness of liquid state or solid state (thawed) water falling from the sky to the ground that has not evaporated, infiltrated or run off. The calculation method is as follows:

Monthly precipitation is the summation of daily precipitation of a month.

Annual precipitation is the summation of 12 months precipitation of a year.

14. Sunshine Hours refer to the actual hours of sun irradiating the earth. The calculation method is the same as that of the precipitation.

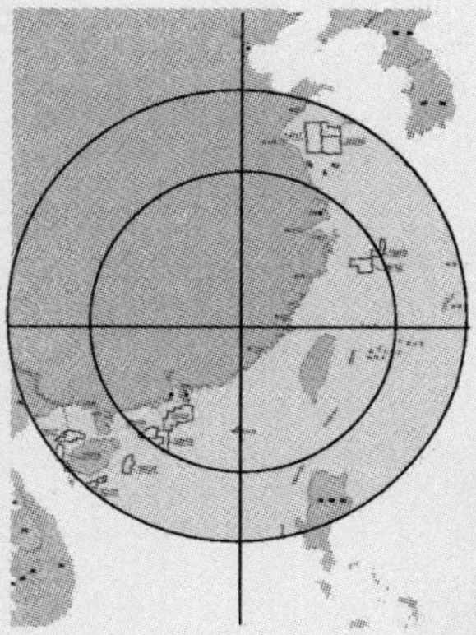

2 海洋经济核算 Marine Economic Accounting

2-1 全国海洋生产总值
National Gross Ocean Product

年 份 Year	海洋生产总值（亿元） Gross Ocean Product (100 million yuan)	第一产业 Primary Industry	第二产业 Secondary Industry	第三产业 Tertiary Industry	海洋生产总值占国内生产总值比重（%） Proportion of the Gross Ocean Product in GDP (%)	海洋生产总值增长速度（%） Growth Rate of the Gross Ocean Product (%)
2001	9 518.4	646.3	4 152.1	4 720.1	8.68	
2002	11 270.5	730.0	4 866.2	5 674.3	9.37	19.8
2003	11 952.3	766.2	5 367.6	5 818.5	8.80	4.2
2004	14 662.0	851.0	6 662.8	7 148.2	9.17	16.9
2005	17 655.6	1 008.9	8 046.9	8 599.8	9.55	16.3
2006	21 592.4	1 228.8	10 217.8	10 145.7	9.98	18.0
2007	25 618.7	1 395.4	12 011.0	12 212.3	9.64	14.8
2008	29 718.0	1 694.3	13 735.3	14 288.4	9.46	9.9
2009	32 161.9	1 857.7	14 926.5	15 377.6	9.31	8.8
2010	39 619.2	2 008.0	18 919.6	18 691.6	9.69	15.3
2011	45 580.4	2 381.9	21 667.6	21 530.8	9.42	10.0
2012	50 172.9	2 670.6	23 450.2	24 052.1	9.39	8.1
2013	54 718.3	3 037.7	24 608.9	27 071.7	9.31	7.8
2014	60 699.1	3 109.5	26 660.0	30 929.6	9.54	7.9

2-2 全国海洋生产总值构成
Composition of National Gross Ocean Product

单位：% (%)

年 份 Year	第一产业 Primary Industry	第二产业 Secondary Industry	第三产业 Tertiary Industry
2001	6.8	43.6	49.6
2002	6.5	43.2	50.3
2003	6.4	44.9	48.7
2004	5.8	45.4	48.8
2005	5.7	45.6	48.7
2006	5.7	47.3	47.0
2007	5.4	46.9	47.7
2008	5.7	46.2	48.1
2009	5.8	46.4	47.8
2010	5.1	47.8	47.2
2011	5.2	47.5	47.2
2012	5.3	46.7	47.9
2013	5.6	45.0	49.5
2014	5.1	43.9	51.0

2-3 海洋及相关产业增加值
Added Values of Marine and Related Industries

单位：亿元 (100 million yuan)

年 份 Year	合 计 Total	海洋产业 Marine Industry	主要海洋产业 Major Marine Industry	海洋科研教育管理服务业 Industries of Marine Scientific Research, Education, Management and Service	海洋相关产业 Ocean-related Industries
2001	9 518.4	5 733.6	3 856.6	1 877.0	3 784.8
2002	11 270.5	6 787.3	4 696.8	2 090.5	4 483.2
2003	11 952.3	7 137.7	4 754.4	2 383.3	4 814.6
2004	14 662.0	8 710.1	5 827.7	2 882.5	5 951.9
2005	17 655.6	10 539.0	7 188.0	3 350.9	7 116.6
2006	21 592.4	12 696.7	8 790.4	3 906.4	8 895.6
2007	25 618.7	15 070.6	10 478.3	4 592.3	10 548.0
2008	29 718.0	17 591.2	12 176.0	5 415.2	12 126.8
2009	32 161.9	18 769.4	12 768.4	6 001.0	13 392.5
2010	39 619.2	22 886.4	16 187.8	6 698.5	16 732.8
2011	45 580.4	26 517.6	18 865.2	7 652.4	19 062.8
2012	50 172.9	29 404.6	20 829.9	8 574.8	20 768.2
2013	54 718.3	32 658.7	22 462.3	10 196.4	22 059.6
2014	60 699.1	36 364.9	25 303.4	11 061.5	24 334.1

2-4 海洋及相关产业增加值构成
Composition of the Added Values of Marine and Related Industries

单位：% (%)

年 份 Year	合 计 Total	海洋产业 Marine Industry	主要海洋产业 Major Marine Industry	海洋科研教育管理服务业 Industries of Marine Scientific Research, Education, Management and Service	海洋相关产业 Ocean-related Industries
2001	100.0	60.2	40.5	19.7	39.8
2002	100.0	60.2	41.7	18.5	39.8
2003	100.0	59.7	39.8	19.9	40.3
2004	100.0	59.4	39.7	19.7	40.6
2005	100.0	59.7	40.7	19.0	40.3
2006	100.0	58.8	40.7	18.1	41.2
2007	100.0	58.8	40.9	17.9	41.2
2008	100.0	59.2	41.0	18.2	40.8
2009	100.0	58.4	39.7	18.7	41.6
2010	100.0	57.8	40.9	16.9	42.2
2011	100.0	58.2	41.4	16.8	41.8
2012	100.0	58.6	41.5	17.1	41.4
2013	100.0	59.7	41.1	18.6	40.3
2014	100.0	59.9	41.7	18.2	40.1

2-5 全国主要海洋产业增加值
Added Values of National Major Marine Industries

主要海洋产业 Major Marine Industry	增加值 (亿元) Added Value (100 million yuan)	比上年增长（%） (按可比价计算) Percentage of Increase over Last Year (%) (at comparable price)
合计 **Total**	**25 303.4**	**8.1**
海洋渔业 Marine Fishery Industry	4 126.6	2.1
海洋油气业 Offshore Oil and Natural Gas Industry	1 530.4	-5.0
海洋矿业 Marine Mining Industry	59.6	11.2
海洋盐业 Sea Salt Industry	68.3	8.8
海洋船舶工业 Marine Shipbuilding Industry	1 395.5	15.5
海洋化工业 Marine Chemical Industry	920.0	15.4
海洋生物医药业 Marine Biomedicine Industry	258.1	7.3
海洋工程建筑业 Marine Engineering Architecture Industry	1 735.0	7.5
海洋电力业 Marine Electric Power Industry	107.7	17.6
海水利用业 Marine Seawater Utilization Industry	12.7	3.7
海洋交通运输业 Maritime Communications and Transportation Industry	5 336.9	9.6
滨海旅游业 Coastal Tourism	9 752.8	11.1

2-6 海洋渔业增加值
Added Value of Marine Fishery Industry

单位：亿元 (100 million yuan)

年 份 Year	增加值 Added Value
2001	966.0
2002	1 091.2
2003	1 145.0
2004	1 271.2
2005	1 507.6
2006	1 672.0
2007	1 906.0
2008	2 228.6
2009	2 440.8
2010	2 851.6
2011	3 202.9
2012	3 560.5
2013	3 997.6
2014	4 126.6

2-7 海洋油气业增加值
Added Value of Offshore Oil and Gas Industry

单位：亿元 (100 million yuan)

年 份 Year	增加值 Added Value
2001	176.8
2002	181.8
2003	257.0
2004	345.1
2005	528.2
2006	668.9
2007	666.9
2008	1 020.5
2009	614.1
2010	1 302.2
2011	1 719.7
2012	1 718.7
2013	1 666.6
2014	1 530.4

2-8　海洋矿业增加值
Added Value of Marine Mining Industry

单位：亿元　　(100 million yuan)

年　份 Year	增加值 Added Value
2001	1.0
2002	1.9
2003	3.1
2004	7.9
2005	8.3
2006	13.4
2007	16.3
2008	35.2
2009	41.6
2010	45.2
2011	53.3
2012	45.1
2013	54.0
2014	59.6

注：自2008年起部分地区统计矿种增加。

Note: The data from 2008 include added kinds of minerals in some regions.

2-9　海洋盐业增加值
Added Value of Marine Salt Industry

单位：亿元　　(100 million yuan)

年　份 Year	增加值 Added Value
2001	32.6
2002	34.2
2003	28.4
2004	39.0
2005	39.1
2006	37.1
2007	39.9
2008	43.6
2009	43.6
2010	65.5
2011	76.8
2012	60.1
2013	63.2
2014	68.3

2-10 海洋船舶工业增加值
Added Value of Marine Shipbuilding Industry

单位：亿元 (100 million yuan)

年 份 Year	增加值 Added Value
2001	109.3
2002	117.4
2003	152.8
2004	204.1
2005	275.5
2006	339.5
2007	524.9
2008	742.6
2009	986.5
2010	1 215.6
2011	1 352.0
2012	1 291.3
2013	1 213.2
2014	1 395.5

2-11 海洋化工业增加值
Added Value of Marine Chemical Industry

单位：亿元 (100 million yuan)

年 份 Year	增加值 Added Value
2001	64.7
2002	77.1
2003	96.3
2004	151.5
2005	153.3
2006	440.4
2007	506.6
2008	416.8
2009	465.3
2010	613.8
2011	695.9
2012	843.0
2013	813.9
2014	920.0

注：自2006年起部分地区统计产品品种增加。

Note: The data from 2006 include added kinds of statistical products in some regions.

2-12　海洋生物医药业增加值
Added Value of Marine Biomedicine Industry

单位：亿元　　(100 million yuan)

年　份 Year	增加值 Added Value
2001	5.7
2002	13.2
2003	16.5
2004	19.0
2005	28.6
2006	34.8
2007	45.4
2008	56.6
2009	52.1
2010	83.8
2011	150.8
2012	184.7
2013	238.7
2014	258.1

2-13　海洋工程建筑业增加值
Added Value of Marine Engineering Architecture

单位：亿元　　(100 million yuan)

年　份 Year	增加值 Added Value
2001	109.2
2002	145.4
2003	192.6
2004	231.8
2005	257.2
2006	423.7
2007	499.7
2008	347.8
2009	672.3
2010	874.2
2011	1 086.8
2012	1 353.8
2013	1 595.5
2014	1 735.0

2-14 海洋电力业增加值
Added Value of Marine Electric Power Industry

单位：亿元 (100 million yuan)

年 份 Year	增加值 Added Value
2001	1.8
2002	2.2
2003	2.8
2004	3.1
2005	3.5
2006	4.4
2007	5.1
2008	11.3
2009	20.8
2010	38.1
2011	59.2
2012	77.3
2013	91.5
2014	107.7

2-15 海水利用业增加值
Added Value of Seawater Utilization Industry

单位：亿元 (100 million yuan)

年 份 Year	增加值 Added Value
2001	1.1
2002	1.3
2003	1.7
2004	2.4
2005	3.0
2006	5.2
2007	6.2
2008	7.4
2009	7.8
2010	8.9
2011	10.4
2012	11.1
2013	11.9
2014	12.7

2-16　海洋交通运输业增加值
Added Value of Marine Communications and Transportation Industry

单位：亿元　　(100 million yuan)

年　份 Year	增加值 Added Value
2001	1 316.4
2002	1 507.4
2003	1 752.5
2004	2 030.7
2005	2 373.3
2006	2 531.4
2007	3 035.6
2008	3 499.3
2009	3 146.6
2010	3 785.8
2011	4 217.5
2012	4 752.6
2013	4 876.5
2014	5 336.9

2-17　滨海旅游业增加值
Added Value of Coastal Tourism

单位：亿元　　(100 million yuan)

年　份 Year	增加值 Added Value
2001	1 072.0
2002	1 523.7
2003	1 105.8
2004	1 522.0
2005	2 010.6
2006	2 619.6
2007	3 225.8
2008	3 766.4
2009	4 277.1
2010	5 303.1
2011	6 239.9
2012	6 931.8
2013	7 839.7
2014	9 752.8

2-18 沿海地区海洋生产总值
Gross Ocean Product by Coastal Regions

地 区 Region	海洋生产总值（亿元） Gross Ocean Product (100 million yuan)	第一产业 Primary Industry	第二产业 Secondary Industry	第三产业 Tertiary Industry	海洋生产总值占沿海地区生产总值比重（%） Proportion of the Gross Ocean Product in the Gross Regional Product (%)
合 计 Total	**60 699.1**	**3 109.5**	**26 660.0**	**30 929.6**	**16.3**
天 津 Tianjin	5 032.2	14.6	3 127.3	1 890.4	32.0
河 北 Hebei	2 051.7	75.2	1 008.3	968.2	7.0
辽 宁 Liaoning	3 917.0	418.7	1 411.0	2 087.3	13.7
上 海 Shanghai	6 249.0	4.3	2 278.4	3 966.2	26.5
江 苏 Jiangsu	5 590.2	316.2	2 894.7	2 379.3	8.6
浙 江 Zhejiang	5 437.7	427.6	2 004.5	3 005.7	13.5
福 建 Fujian	5 980.2	480.8	2 299.3	3 200.2	24.9
山 东 Shandong	11 288.0	794.5	5 089.0	5 404.5	19.0
广 东 Guangdong	13 229.8	201.0	5 993.9	7 034.9	19.5
广 西 Guangxi	1 021.2	175.9	373.5	471.7	6.5
海 南 Hainan	902.1	200.8	180.1	521.2	25.8

2-19 沿海地区海洋生产总值构成
Composition of Gross Ocean Product by Coastal Regions

单位：%　　　　(%)

地　区 Region	海洋生产总值 Gross Ocean Product	第一产业 Primary Industry	第二产业 Secondary Industry	第三产业 Tertiary Industry
合　计 Total	**100.0**	**5.1**	**43.9**	**51.0**
天　津 Tianjin	100.0	0.3	62.1	37.6
河　北 Hebei	100.0	3.7	49.1	47.2
辽　宁 Liaoning	100.0	10.7	36.0	53.3
上　海 Shanghai	100.0	0.1	36.5	63.5
江　苏 Jiangsu	100.0	5.7	51.8	42.6
浙　江 Zhejiang	100.0	7.9	36.9	55.3
福　建 Fujian	100.0	8.0	38.4	53.5
山　东 Shandong	100.0	7.0	45.1	47.9
广　东 Guangdong	100.0	1.5	45.3	53.2
广　西 Guangxi	100.0	17.2	36.6	46.2
海　南 Hainan	100.0	22.3	20.0	57.8

2-20 沿海地区海洋及相关产业增加值
Added Values of Marine and Related Industries by Coastal Regions

单位：亿元 (100 million yuan)

地 区 Region	合 计 Total	海洋产业 Marine Industry	主要海洋产业 Major Marine Industry	海洋科研教育管理服务业 Industries of Marine Scientific Research, Education, Management and Service	海洋相关产业 Ocean-related Industries
合 计 Total	**60 699.1**	**36 364.9**	**25 303.4**	**11 061.5**	**24 334.1**
天 津 Tianjin	5 032.2	2 788.8	2 533.0	255.8	2 243.5
河 北 Hebei	2 051.7	1 136.7	1 046.6	90.1	915.1
辽 宁 Liaoning	3 917.0	2 507.2	1 927.8	579.4	1 409.8
上 海 Shanghai	6 249.0	3 756.1	2 081.1	1 675.0	2 492.9
江 苏 Jiangsu	5 590.2	3 152.0	2 264.9	887.2	2 438.2
浙 江 Zhejiang	5 437.7	3 335.8	2 263.2	1 072.6	2 101.9
福 建 Fujian	5 980.2	3 407.9	2 617.6	790.4	2 572.3
山 东 Shandong	11 288.0	6 832.3	4 835.0	1 997.3	4 455.7
广 东 Guangdong	13 229.8	8 167.6	4 763.7	3 403.9	5 062.1
广 西 Guangxi	1 021.1	639.4	539.5	100.0	381.7
海 南 Hainan	902.1	641.2	431.1	210.1	260.9

2-21 沿海地区海洋及相关产业增加值构成
Composition of the Added Values of Marine and Related Industries by Coastal Regions

单位：% (%)

地区 Region	合计 Total	海洋产业 Marine Industry	主要海洋产业 Major Marine Industry	海洋科研教育管理服务业 Industries of Marine Scientific Research, Education, Management and Service	海洋相关产业 Ocean-related Industries
合计 Total	**100.0**	**59.9**	**41.7**	**18.2**	**40.1**
天津 Tianjin	100.0	55.4	50.3	5.1	44.6
河北 Hebei	100.0	55.4	51.0	4.4	44.6
辽宁 Liaoning	100.0	64.0	49.2	14.8	36.0
上海 Shanghai	100.0	60.1	33.3	26.8	39.9
江苏 Jiangsu	100.0	56.4	40.5	15.9	43.6
浙江 Zhejiang	100.0	61.3	41.6	19.7	38.7
福建 Fujian	100.0	57.0	43.8	13.2	43.0
山东 Shandong	100.0	60.5	42.8	17.7	39.5
广东 Guangdong	100.0	61.7	36.0	25.7	38.3
广西 Guangxi	100.0	62.6	52.8	9.8	37.4
海南 Hainan	100.0	71.1	47.8	23.3	28.9

主要统计指标解释

1. 海洋经济 是开发、利用和保护海洋的各类产业活动以及与之相关联活动的总和。

2. 海洋生产总值 是海洋经济生产总值的简称，指按市场价格计算的沿海地区常住单位在一定时期内海洋经济活动的最终成果，是海洋产业和海洋相关产业增加值之和。

3. 海洋产业 是开发、利用和保护海洋所进行的生产和服务活动，包括海洋渔业、海洋油气业、海洋矿业、海洋盐业、海洋化工业、海洋生物医药业、海洋电力业、海水利用业、海洋船舶工业、海洋工程建筑业、海洋交通运输业、滨海旅游业等主要海洋产业以及海洋科研教育管理服务业。

4. 海洋科研教育管理服务业 是开发、利用和保护海洋过程中所进行的科研、教育、管理及服务等活动，包括海洋信息服务业、海洋环境监测预报服务、海洋保险与社会保障业、海洋科学研究、海洋技术服务业、海洋地质勘查业、海洋环境保护业、海洋教育、海洋管理、海洋社会团体与国际组织等。

5. 海洋相关产业 是指以各种投入产出为联系纽带，与主要海洋产业构成技术经济联系的上下游产业，涉及海洋农林业、海洋设备制造业、涉海产品及材料制造业、涉海建筑与安装业、海洋批发与零售业、涉海服务业等。

6. 海洋三次产业 我国的海洋三次产业划分如下：

海洋第一产业：是指海洋渔业中的海洋水产品、海洋渔业服务业，以及海洋相关产业中属于第一产业范畴的部门。

海洋第二产业：是指海洋渔业中海洋水产品加工、海洋油气业、海洋矿业、海洋盐业、海洋化工业、海洋生物医药业、海洋电力业、海水利用业、海洋船舶工业、海洋工程建筑业，以及海洋相关产业中属于第二产业范畴的部门。

海洋第三产业：是指除海洋第一、第二产业以外的其他行业。第三产业包括：海洋交通运输业、滨海旅游业、海洋科研教育管理服务业，以及海洋相关产业中属于第三产业范畴的部门。

7. 海洋渔业 包括海水养殖、海洋捕捞、海洋渔业服务业和海洋水产品加工等活动。

8. 海洋油气业 是指在海洋中勘探、开采、输送、加工原油和天然气的生产活动。

9. 海洋矿业 包括海滨砂矿、海滨土砂石与煤矿及深海矿物等的采选活动。

10. 海洋盐业 是指利用海水生产以氯化钠为主要成分的盐产品的活动，包括采盐和盐加工。

11. 海洋船舶工业 是指以金属或非金属为主要材料，制造海洋船舶、海上固定及浮动装置的活动，以及对海洋船舶的修理及拆卸活动。

12. 海洋化工业 包括海盐化工、海水化工、海藻化工及海洋石油化工的化工产品生产活动。

13. 海洋生物医药业 是指以海洋生物为原料或提取有效成分，进行海洋药品与海洋保健品的生产加工及制造活动。

14. 海洋工程建筑业 是指在海上、海底和海岸所进行的用于海洋生产、交通、娱乐、防护等用途的建筑工程施工及其准备活动；包括海港建筑、滨海电站建筑、海岸堤坝建筑、海洋隧道桥梁建筑、海上油气田陆地终端及处理设施建造、海底线路管道和设备安装，不包括各部门、各地区的房屋建筑及房屋装修工程。

15. 海洋电力业 是指在沿海地区利用海洋能、海洋风能进行的电力生产活动。不包括沿海地区的火力发电和核力发电。

16. 海水利用业 是指对海水的直接利用和海水淡化活动，包括利用海水进行淡水生产和将海水应用于工业冷却用水和城市生活用水、消防用水等活动，不包括海水化学资源综合利用活动。

17. 海洋交通运输业 是指以船舶为主要工具从事海洋运输以及为海洋运输提供服务的活动，包括远洋旅客运输、沿海旅客运输、远洋货物运输、沿海货物运输、水上运输辅助活动、管道运输业、装卸搬运及其他运输服务活动。

18. 滨海旅游业 是指以海岸带、海岛及海洋各种自然景观、人文景观为依托的旅游经营、服务活动，主要包括：海洋观光游览、休闲娱乐、度假住宿、体育运动等活动。

Explanatory Notes on Main Statistical Indicators

1. Marine Economy is the summation of various types of industrial activities for developing, utilizing and protecting the ocean as well as the activities associated with there.

2. Gross Ocean Product is the short form of the gross output value of ocean economy, referring to the final result of marine economic activities of the permanent units in the coastal region within a given period calculated at the market price, and the sum total of the added values of the marine industries as the ocean-related industries.

3. Marine industry refers to the production as service activities for developing, utilizing and protecting the ocean, including major marine industries such as offshore oil and gas industry, marine mining industry, marine salt industry, marine chemical industry, marine biomedicine industry, marine electric power industry, seawater utilization industry, marine shipbuilding industry, marine engineering construction industry, marine communications and transportation industry, coastal tourism etc. as well as marine scientific research, education, management and service.

4. Marine Scientific Research, Education, Management and Service refer to the activities of scientific research, education, management and service carried out in the process of developing, utilizing and protecting the ocean, including marine information service industry, marine environment monitoring and forecasting service, marine insurance and social security industry, marine scientific research, marine technological service industry, ocean geological prospecting industry, marine environmental protection industry, marine education, marine management, marine social organization and international organizations etc.

5. Ocean-Related Industry refers to the lower and upper reaches enterprises that form a technical and economic link with the major marine industries, with various inputs and outputs as ties, involving marine agriculture and forestry, marine equipment manufacturing, ocean-related building and installation industry, marine wholesale and retail industry, ocean-related service industry etc.

6. Marine Three Industries Chinese marine three industries are divided as follows:

Marine primary industry: refers to the marine aquatic products, marine fishery service industry in the marine fishery as well as the sectors belonging to the primary industry category in the ocean-related industries.

Marine secondary industry: refers to the marine aquatic products processing industry in the marine fishery, offshore oil as gas industry, marine mining industry, marine salt industry, marine chemical industry, marine biomedicine industry, marine electric power industry, seawater utilization industry, marine shipbuilding industry, marine engineering construction industry, as well as the sectors belonging to the category of secondary industry in the ocean-related industries.

Marine tertiary Industry: refers to the industries other than the marine primary and secondary industries, including marine communications and transportation industry, coastal tourism, marine scientific research, education, management and service industry as well as the sectors belonging to the category of tertiary industry in the ocean-related industries.

7. Marine Fishery includes mariculture, marine fishing, marine fishery service industry and marine aquatic products processing, etc.

8. Offshore Oil and Gas Industry refers to the production activities of exploring, exploiting, transporting and processing crude oil and natural gas in the ocean.

9. Ocean Mining Industry includes the activities of extracting and dressing beach placers, beach soil and sand, and coal mining and deep-sea mining, etc.

10. Marine Salt Industry refers to the activity of producing the salt products with the sodium chloride as the main component by utilizing seawater, including salt extracting and processing.

11. Shipbuilding Industry refers to the activity of building ocean vessels, offshore fixed and floating equipment with metals or non-metals as main materials as well as repairing and dismantling ocean vessels.

12. Marine Chemical Industry includes the production activities of chemical products of sea salt, seawater, sea algal and marine petroleum chemical industries.

13. Marine Biomedicine Industry refers to the production, processing and manufacturing activities of marine medicines and marine health care products by using marine organisms as raw materials or extracting useful components therefrom.

14. Marine Engineering Building Industry refers to the architectural projects construction and its preparations in the sea, at the sea bottom and seacoast for such uses as marine production, transportation, recreation, protection, etc., including constructions of seaports, coastal power stations, coastal dykes, marine tunnels and bridges, land terminals of offshore oil and gas fields as well as building of processing facilities, and installation of submarine pipelines and equipment, but not the projects of house building and renovation.

15. Marine Electric Power Industry refers to the activities of generating electric power in the coastal region by making use of ocean energies and ocean wind energy. It does not include the thermal and nuclear power generation in the coastal area.

16. Seawater Utilization Industry refers to the activities of the direct use of sea water and the seawater desalination, including those of carrying out the production of desalination and applying the seawater as water for industrial cooling, urban domestic water, water for fire fighting etc., but not the activity of the multipurpose use of seawater chemical resources.

17. Marine Communications and Transportation Industry refers to the activities of carrying out and serving the sea transportations with vessels as main vehicles, including ocean-going passengers transportation, coastal passengers transportation, ocean-going cargo transportation, coastal cargo transportation, auxiliary activities of water transportation, pipeline transportation, loading, unloading and transport as well as other transportation service activities.

18. Coastal Tourism refers to the tourist business and service activities with the backing of coastal zone, sea islands as well as a variety of natural and human landscapes of the ocean, mainly including marine sightseeing, living a life of leisure and recreation, going on vocation and getting accommodation, sports, etc.

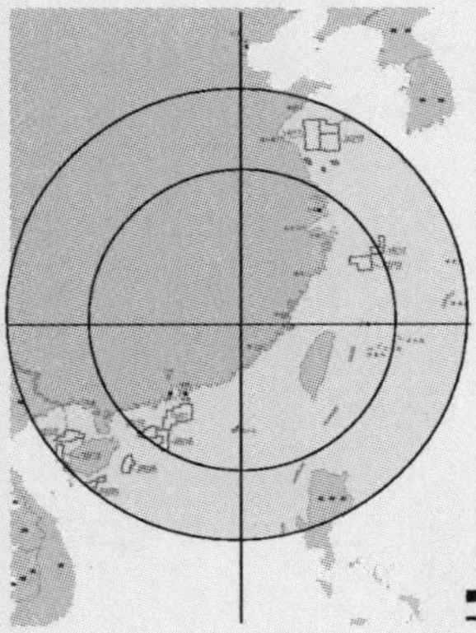

3 主要海洋产业活动

Major Marine Industrial Activities

3-1 全国海洋捕捞和海水养殖产量
National Marine Catches and Mariculture Production

单位：吨 (t)

项　目　Item	2012	2013	2014
远洋捕捞产量 Deep-sea fishing production	**1 223 441**	**1 351 978**	**2 027 318**
海洋捕捞产量 Marine Catches	**12 671 891**	**12 643 822**	**12 808 371**
按品种分 By Species			
鱼类 Fish	8 758 466	8 717 638	8 807 901
甲壳类 Crustacea	2 207 391	2 285 476	2 395 699
贝类 Shellfish	563 422	547 556	551 607
藻类 Algae	25 728	28 036	24 299
头足类 Cephalopoda	698 909	664 285	676 715
其他 Others	417 975	400 831	352 150
海水养殖产量 Mariculture Production	**16 438 105**	**17 392 453**	**18 126 481**
按品种分 By Species			
鱼类 Fish	1 028 399	1 123 576	1 189 667
甲壳类 Crustacea	1 249 554	1 340 218	1 433 763
贝类 Shellfish	12 084 393	12 728 037	13 165 511
藻类 Algae	1 764 684	1 856 804	2 004 576
其他 Others	311 075	343 818	332 964

3-2 沿海地区海洋捕捞和海水养殖产量
Marine Catches and Mariculture Production by Coastal Regions

单位：吨 (t)

地 区 Region	海洋捕捞产量 Marine Catches	远洋捕捞产量 Deep-Sea Fishing Production	海水养殖产量 Mariculture Production
全国总计 National Total	**12 808 371**	**2 027 318**	**18 126 481**
天 津 Tianjin	45 548	20 046	11 627
河 北 Hebei	239 595		491 999
辽 宁 Liaoning	1 076 005	330 295	2 890 525
上 海 Shanghai	19 945	149 649	
江 苏 Jiangsu	547 952	19 907	935 947
浙 江 Zhejiang	3 242 724	532 666	897 940
福 建 Fujian	1 975 062	264 487	3 794 298
山 东 Shandong	2 297 194	365 042	4 799 107
广 东 Guangdong	1 493 656	68 370	2 943 981
广 西 Guangxi	650 599	2 787	1 090 975
海 南 Hainan	1 220 091		270 082

注：远洋捕捞产量全国总计包括北京13 222吨，中农发集团260 847吨。

Note:Data for the national total deep-sea fishing production include 13 222 tons from Beijing and 260 847 tons from the China National Agricultural Development Group Co., Ltd..

3-3 沿海地区海洋原油产量
Output of Offshore Crude Oil by Coastal Regions

单位：万吨 (10 000 t)

地 区 Region	2012	2013	2014
合 计 Total	**4 444.79**	**4 541.09**	**4 613.95**
天 津 Tianjin	2 680.34	2 634.68	2 674.09
河 北 Hebei	237.77	239.87	237.74
辽 宁 Liaoning	14.25	11.29	48.27
上 海 Shanghai	14.68	18.65	19.53
山 东 Shandong	275.00	291.30	300.30
广 东 Guangdong	1 222.75	1 345.30	1 334.02

3-4 沿海地区海洋天然气产量
Output of Offshore Natural Gas by Coastal Regions

单位：万立方米 (10 000 m^3)

地 区 Region	2012	2013	2014
合 计 Total	**1 228 188**	**1 176 455**	**1 308 899**
天 津 Tianjin	246 705	261 682	282 592
河 北 Hebei	55 570	67 201	85 250
辽 宁 Liaoning	1 580	1 410	1 881
上 海 Shanghai	88 044	79 702	88 071
山 东 Shandong	12 521	12 911	12 900
广 东 Guangdong	823 768	753 549	838 205

3-5 沿海地区海洋矿业产量
Output of Marine Mining Industry by Coastal Regions

单位：万吨 (10 000t)

地 区 Region	产 量 Output		
	2012	2013	2014
合 计 Total	**4 351.4**	**4 434.2**	**4 727.4**
浙 江 Zhejiang	2 727.0	2 134.4	2 243.7
福 建 Fujian	264.6	299.5	316.7
山 东 Shandong	1 018.4	1 318.4	1 507.0
广 西 Guangxi	35.7	481.0	454.4
海 南 Hainan	305.7	200.9	205.6

3-6　沿海地区海盐产量
Sea Salt Production by Coastal Regions

单位：万吨　　(10 000 t)

地　区 Region	海盐产量 Output of Sea Salt		
	2012	2013	2014
合　计 Total	**2 986.4**	**2 681.1**	**3 085.3**
天　津 Tianjin	170.0	152.2	161.1
河　北 Hebei	334.7	287.6	336.3
辽　宁 Liaoning	117.4	98.3	128.7
江　苏 Jiangsu	78.1	77.6	81.4
浙　江 Zhejiang	10.9	15.9	8.9
福　建 Fujian	27.0	29.2	38.1
山　东 Shandong	2 219.1	1 989.9	2 316.6
广　东 Guangdong	8.6	7.8	8.4
广　西 Guangxi	16.4	16.1	0.6
海　南 Hainan	4.2	6.6	5.3

3-7 沿海地区海洋化工产品产量
Output of Marine Chemical Products by Coastal Regions

单位：吨 (t)

地区 Region	产品产量[①] Output		
	2012	2013	2014
合计 Total	**17 450 267**	**19 095 811**	**26 382 926**
天津 Tianjin	1 610 000	1 641 323	4 586 402
河北 Hebei	1 046 405[②]	7 942[②]	2 037 150[②]
辽宁 Liaoning	1 020 284	1 021 941	1 162 236
江苏 Jiangsu	2 090 567	1 933 766	2 523 815
浙江 Zhejiang	1 001 170	1 022 884	1 086 436
福建 Fujian	1 462 106	1 652 964	2 146 419
山东 Shandong	8 414 735	10 919 468	11 640 468
广东 Guangdong	805 000[②]	895 523[②]	1 200 000[②]

注：①数据为沿海地区部分海洋化工企业产品汇总数据；②为中国盐业总公司数据。

Note: ① The data are collected from the products of part of the chemical enterprises in the coastal region.
② The data are from the China Salt Industry Corporation.

3-8 沿海地区海洋生物医药产品产量
Production of the Marine Biomedicine Industry by Coastal Regions

产品名称 Name	计量单位 Unit	产品产量* Output
海参肽胶囊 Sea Cucumber Peptide Capsule	吨 t	270.69
螺旋藻粉 Spirulina Powder	吨 t	5 046.00
螺旋藻胶囊 Spirulina Capsule	万粒 10 000 pellets	9 550.00
海藻酸钠 Sodium Algrnic Acid	万吨 10 000 tons	1.96
鲨鱼肝油胶丸 Shark Liver Oil Pill	万粒 10 000 pellets	3 668.70
金枪鱼油胶丸 Tuna Oil Pill	万粒 10 000 pellets	877.20
蚝贝钙片 Oyster Shell Calcium Tablets	万片 10 000 pills	39 273.64
鱼肝油 Cod-Liver Oil	万瓶 10 000 bottles	508.95
鱼肝油乳 Cod-Liver Oil Milk	万瓶 10 000 bottles	221.84
鱼肝油系列产品 Cod-Liver Oil Series Products	箱 box	2 200.00
鲨鱼硫酸软骨素 Shark Sulphate Chondroitin	吨 t	552.28
海蛇痹宁胶囊 Sea Snake Anti-rheumatic Capsule	万粒 10 000 pellets	382.00
维生素AD胶丸 Vitamin AD Capsule	万粒 10 000 pellets	77 963.40
维生素AD滴剂 Vitamin AD Drops	万瓶 10 000 bottles	50 348.82
维生素E胶丸 Vitamin E Capsule	万粒 10 000 pellets	44 755.27
琼脂 Agar	吨 t	2 500.00
海珠喘息定片 Haizhu Methoxyphenamine Pill	万片 10 000 pills	18 733.91
珍珠粉末 Pearl Powder	万支 10 000 bottles	299.50
金牡感冒片 Jinmu Cold Cure Pill	万片 10 000 pills	1 728.74
多烯酸乙酯胶囊 Ethyl Polyenoic Acid Capsule	万粒 10 000 pellets	220.73
氨糖美辛肠溶片 Glucosamine Indometacin Enteric-coated Tablets	万片 10 000 pills	759.35
去甲斑蝥素片 Demethylcantharidin Tablets	万片 10 000 pills	112.00

3-8 续表 continued

产品名称 Name	计量单位 Unit	产品产量* Output
浓缩鱼油 Concentrated Fish Oil	吨 t	1 500.00
甘糖酯 Mannose Easter	万片 10 000 pills	322.30
珠珀惊风散 Zhubo Convulsions Powder	万袋 10 000 bags	468.92
甲壳素 Crustaceoxin	万盒 10 000 cases	22.50
虾青素Astaxanthin	吨 t	474.09
脑元神软胶囊 Naoyuanshen Soft Capsule	万粒 10 000 pellets	307.18
海麟舒肝胶囊 Hailin Liver-Soothing Capsule	万粒 10 000 pellets	27.20
海威口服液 Haiwei Oral Liquid	吨 t	518.00
伊可新 Vitamin A and D Drops	吨 t	250.00
珍珠胎囊口服液 Pearl Capsule Oral Liquid	万支 10 000 bottles	420.00
麝珠明目滴眼液 Shezhu Eye-brightening Drops	万瓶 10 000 bottles	198.00
卡拉胶 Carrageenan	吨 t	9 542.00
角鲨烯胶囊 Houndfish Alkene Capsule	万粒 10 000 pellets	546.39
鱼蛋白粉 Fish Albumen Powder	万罐 10 000 can	0.82
胶原蛋白 Collagen Albumen	吨 t	327.09
鱼油微囊 Fish Oil Capsule	吨 t	3 600.00
鱼胶原蛋白肽 Fish Collagen Peptide	万粒 10 000 pellets	112.09
八宝惊风散 Babao Infantile Convulsions Powder	万瓶 10 001 bottles	729.64
碘 Iodine	吨 t	3 710.00
润科DHA粉剂、油剂 Runke DHA Powder, oil	吨 t	1 650.00
润科ARA粉剂、油剂 Runke ARA Powder,oil	吨 t	1 100.00
鳗钙 Eel-calcium	万盒 10 000 cases	280.00
肤疹宁软膏 Rash Cream	万支 10 000 bottles	5.00
康体通胶囊 Kangtitong Capsule	吨 t	28.00
鲨鱼软骨罐头 Shark Cartilage Can	罐 can	60 000.00

注：数据为沿海地区部分海洋生物医药产品汇总数据。

Note: The data are collected from the products of part of the marine biomedicine enterprices in the coastal region.

3-9 沿海地区海洋修造船完工量
Production of the Marine Shipbuilding Industry by Coastal Regions

部门和地区 Sector and Region	修船完工量（艘） Ships Repaired (unit)	造船完工量 Ships Built	
		艘 (unit)	万综合吨 Comprehensive Tonnages (10 000 t)
总 计 Total	**12 404**	**3 315**	**3 950.7**
其 中: Including:			
中船工业集团公司 CSSC	613	174	1 166.9
中船重工集团公司 CSIC	640	74	563.3
按地区分: By Regions:			
天 津 Tianjin	209	17	85.9
河 北 Hebei	1 802	10	15.9
辽 宁 Liaoning	155	59	485.0 ①
上 海 Shanghai	1 199	102	956.4
江 苏 Jiangsu	456	329	1 256.4
浙 江 Zhejiang	4 471	826	610.9
福 建 Fujian	2 370	1 027	90.8
山 东 Shandong	1 498	746	373.5
广 东 Guangdong	139	24	73.9
广 西 Guangxi	85	20	0.8
海 南 Hainan	20	155	1.2

注：①表示单位为万载重吨。

Note: ①the unit is 10 thousand dead weight tonnage.

3-10 沿海地区海洋货物运输量和周转量
Volume of Maritime Goods Transported and Turnover by Coastal Regions

单位：万吨，亿吨・千米 (10 000 t, 100 million t-km)

地区 Region	货运量 Volume of Goods Transported	沿海 Coastal	远洋 Oceangoing	货物周转量 Volume of Goods Turnover	沿海 Coastal	远洋 Oceangoing
全国总计 National Total	**250 814**	**176 492**	**74 322**	**78 427**	**22 778**	**55 647**
天津 Tianjin	9 749	8 111	1 638	2 734	1 252	1 482
河北 Hebei	4 041	2 726	1 315	1 482	443	1 039
辽宁 Liaoning	13 810	6 303	7 507	7 980	814	7 166
上海 Shanghai	44 231	27 690	16 541	18 274	3 787	14 487
江苏 Jiangsu	23 725	16 570	7 155	6 582	1 894	4 687
浙江 Zhejiang	51 724	47 855	3 869	7 603	5 386	2 216
福建 Fujian	22 186	20 098	2 088	3 634	3 041	593
山东 Shandong	9 226	8 365	861	1 009	602	407
广东 Guangdong	39 328	23 262	16 066	10 771	3 813	6 958
广西 Guangxi	5 211	4 715	496	680	654	26
海南 Hainan	11 433	10 485	948	1 334	1 055	279
其他 Others	16 150	312	15 838	16 344	37	16 307

3-11 沿海地区海洋旅客运输量和周转量
Volume of Maritime Passenger Traffic and Turnover by Coastal Regions

单位：万人，亿人·千米 (10 000 persons, 100 million person-km)

地区 Region	客运量 Passenger Traffic	沿海 Coastal	远洋 Oceangoing	旅客周转量 Passenger Turnover Volume	沿海 Coastal	远洋 Oceangoing
全国总计 National Total	**10 521**	**9 394**	**1 127**	**41.1**	**27.9**	**13.2**
天津 Tianjin						
河北 Hebei	4		4	0.3		0.3
辽宁 Liaoning	542	527	15	6.5	5.8	0.7
上海 Shanghai	369	368	1	1.1	0.9	0.2
江苏 Jiangsu	33	24	9	0.7	0.0	0.7
浙江 Zhejiang	2 636	2 636		4.6	4.6	
福建 Fujian	1 503	1 420	83	2.4	1.8	0.5
山东 Shandong	1 539	1 429	110	11.6	7.1	4.5
广东 Guangdong	2 167	1 262	905	9.5	3.3	6.2
广西 Guangxi	229	229		1.2	1.2	0.0
海南 Hainan	1 499	1 499		3.3	3.3	

3-12 沿海港口客货吞吐量
Volume of Passenger and Freight Handled at Coastal Seaports

单位：万吨，万人 (10 000 t, 10 000 persons)

地区 Region	货物吞吐量 Cargo Handled	#外贸 Foreign Trade	旅客吞吐量 Passenger Leaving & Arriving	#离港 Leaving
合计 Total	**803 307**	**326 666**	**8 062**	**4 038**
天津 Tianjin	54 002	29 493	33	16
河北 Hebei	95 029	29 330	4	2
辽宁 Liaoning	103 658	22 157	608	311
上海 Shanghai	66 954	38 232	178	90
江苏 Jiangsu	25 744	12 406	9	5
浙江 Zhejiang	108 177	44 042	705	351
福建 Fujian	49 166	20 988	980	489
山东 Shandong	128 593	66 263	1 321	623
广东 Guangdong	137 631	48 385	2 843	1 445
广西 Guangxi	20 189	12 773	20	10
海南 Hainan	14 164	2 597	1 361	696

3-13 沿海地区水路国际标准集装箱运量
Volume of International Standardized Containers Traffic by Coastal Regions

单位：万标准箱，万吨 (10 000 TEU, 10 000 t)

地区 Region	2012		2013		2014	
	箱数 Number of Containers	重量 Weight	箱数 Number of Containers	重量 Weight	箱数 Number of Containers	重量 Weight
合计 Total	**4 357**	**52 299**	**4 646**	**54 766**	**4 931**	**59 739**
天津 Tianjin	6	62	9	168	28	524
河北 Hebei	1	20	1	18	6	91
辽宁 Liaoning	42	473	41	427	42	463
上海 Shanghai	2 010	26 378	1 817	23 722	1 901	25 242
江苏 Jiangsu	479	4 063	508	4 218	475	3 850
浙江 Zhejiang	220	2 948	239	3 291	272	3 671
福建 Fujian	302	4 373	345	5 419	440	6 868
山东 Shandong	139	1 927	120	1 414	117	1 287
广东 Guangdong	786	6 618	975	7 244	982	7 462
广西 Guangxi	210	2 598	200	3 080	259	4 410
海南 Hainan	163	2 839	111	1 874	124	2 949
其他 other			280	3 890	285	2 922

3-14 沿海港口国际标准集装箱吞吐量

International Standardized Containers Handled at Coastal Seaports

单位：万标准箱，万吨 (10 000TEU, 10 000 t)

地 区 Region	2012		2013		2014	
	箱 数 Number of Containers	重 量 Weight	箱 数 Number of Containers	重 量 Weight	箱 数 Number of Containers	重 量 Weight
合 计 Total	**15 797**	**175 986**	**16 967**	**194 694**	**18 179**	**210 056**
天 津 Tianjin	1 230	13 442	1 301	15 216	1 406	15 904
河 北 Hebei	90	1 351	135	2 051	184	2 723
辽 宁 Liaoning	1 514	24 733	1 798	28 381	1 860	30 834
上 海 Shanghai	3 253	32 480	3 362	34 243	3 529	35 335
江 苏 Jiangsu	504	5 012	554	5 536	511	5 074
浙 江 Zhejiang	1 759	17 811	1 910	19 659	2 136	22 224
福 建 Fujian	1 073	13 308	1 169	14 890	1 271	16 502
山 东 Shandong	1 899	19 912	2 076	22 758	2 256	24 845
广 东 Guangdong	4 256	44 495	4 420	47 947	4 752	51 871
广 西 Guangxi	82	1 358	100	1 703	112	1 960
海 南 Hainan	137	2 084	142	2 310	162	2 784

3-15 沿海城市国内旅游人数
Domestic Visitors by Coastal Cities

单位：万人·次 (10 000 person-times)

城市	City	2011	2012	2013
合 计	**Total**	**135 467**	**117 625**	**131 226**
天 津	**Tianjin**	10 605		
河 北	**Hebei**	4 790	5 977	6 304
唐 山	Tangshan	2 001	2 454	2 770
秦皇岛	Qinhuangdao	2 101	2 313	2 565
沧 州	Cangzhou	688	1 210	969
辽 宁	**Liaoning**	13 586	15 225	16 898
大 连	Dalian	4 261	4 687	5 231
丹 东	Dandong	2 646	2 990	2 990
锦 州	Jinzhou	1 704	1 926	3 211
营 口	Yingkou	1 279	1 446	1 620
盘 锦	Panjin	1 959	2 214	1 912
葫芦岛	Huludao	1 736	1 962	1 934
上 海	**Shanghai**	23 079		
江 苏	**Jiangsu**	5 090	5 839	6 606
南 通	Nantong	2 109	2 408	2 716
连云港	Lianyungang	1 656	1 894	2 136
盐 城	Yancheng	1 325	1 537	1 754
浙 江	**Zhejiang**	30 547	35 048	39 788
杭 州	Hangzhou	7 181	8 237	9 409
宁 波	Ningbo	5 181	5 748	6 226
温 州	Wenzhou	4 123	4 887	5 677
嘉 兴	Jiaxing	3 536	4 101	4 660
绍 兴	Shaoxing	4 128	4 866	5 614
舟 山	Zhoushan	2 433	2 740	3 036
台 州	Taizhou	3 965	4 469	5 166
福 建	**Fujian**	9 280	11 665	12 709
福 州	Fuzhou	2 680	3 107	3 446
厦 门	Xiamen	2 569	2 979	3 411
莆 田	Putian	1 165	1 112	1 590
泉 州	Quanzhou	1 083	2 170	1 480
漳 州	Zhangzhou	951	1 348	1 435
宁 德	Ningde	832	949	1 347

3-15 续表 continued

城　市　City		2011	2012	2013
山　东	**Shandong**	18 807	21 660	24 114
青　岛	Qingdao	4 956	5 591	6 166
东　营	Dongying	775	938	1 098
烟　台	Yantai	3 863	4 450	4 952
潍　坊	Weifang	3 603	4 221	4 702
威　海	Weihai	2 372	2 669	2 953
日　照	Rizhao	2 426	2 795	3 125
滨　州	Binzhou	812	996	1 118
广　东	**Guangdong**	15 537	17 411	19 338
广　州	Guangzhou	3 816	4 017	4 274
深　圳	Shenzhen	2 628	2 941	3 352
珠　海	Zhuhai	1 215	1 299	1 309
汕　头	Shantou	890	1 026	1 140
江　门	Jiangmen	1 013	1 120	1 240
湛　江	Zhanjiang	909	1 247	1 392
茂　名	Maoming	360	399	426
惠　州	Huizhou	1 014	1 122	1 295
汕　尾	Shanwei	439	518	582
阳　江	Yangjiang	456	572	700
东　莞	Dongguan	1 400	1 432	1 481
中　山	Zhongshan	605	744	808
潮　州	Chaozhou	361	438	529
揭　阳	Jieyang	432	536	810
广　西	**Guangxi**	2 347	2 811	3 260
北　海	Beihai	1 101	1 311	1 521
防城港	Fangchenggang	676	807	965
钦　州	Qinzhou	570	693	774
海　南	**Hainan**	1 799	1 989	2 209
海　口	Haikou	831	935	1 029
三　亚	Sanya	968	1 054	1 180

注：数据来源于《中国区域经济统计年鉴》（2014）。

Note：The data come from *China Statistical Yearbook For Regional Economy (2014)*.

3-16 主要沿海城市接待入境游客人数
Number of Inbound Tourists Received by Major Coastal Cities

单位：人·次 (person-time)

城　市	City	2012	2013	2014
天　津	Tianjin	737 481	758 594	766 326
秦皇岛	Qinhuangdao	286 401	188 041	184 355
大　连	Dalian	1 284 176	734 605	965 615
上　海	Shanghai	6 512 347	6 140 911	6 396 150
南　通	Nantong	440 788	216 944	187 185
连云港	Lianyungang	144 684	24 228	22 972
杭　州	Hangzhou	3 311 225	1 060 360	984 701
宁　波	Ningbo	1 162 088	426 258	520 641
温　州	Wenzhou	575 397	290 335	479 898
福　州	Fuzhou	851 328	568 853	569 487
厦　门	Xiamen	2 124 163	1 072 263	1 170 058
泉　州	Quanzhou	951 424	622 476	675 944
漳　州	Zhangzhou	330 669	238 194	270 336
青　岛	Qingdao	1 270 076	782 632	836 605
烟　台	Yantai	530 184	328 872	358 748
威　海	Weihai	456 594	278 003	300 744
广　州	Guangzhou	7 866 031	7 681 966	7 832 990
深　圳	Shenzhen	12 064 451	12 148 917	11 825 916
珠　海	Zhuhai	2 975 791	2 632 331	2 913 367
汕　头	Shantou	147 800	155 222	176 500
湛　江	Zhanjiang	147 400	190 247	225 500
中　山	Zhongshan	558 573	537 743	602 050
北　海	Beihai	98 759	80 026	74 512
海　口	Haikou	179 741	156 632	136 910
三　亚	Sanya	481 437	481 851	388 637

3-17 主要沿海城市接待入境游客情况
Breakdown of Inbound Tourists Received by Major Coastal Cities

单位：人·次，人·天　　　　(person-time, night)

城市 City	外国人 Foreigners		香港同胞 Hong Kong	
	人次数 Arrivals	人天数 Nights	人次数 Arrivals	人天数 Nights
天　津 Tianjin	674 931	10 876 790	43 829	1 139 606
秦皇岛 Qinhuangdao	172 102	785 301	5 993	21 743
大　连 Dalian	823 598	1 580 829	66 317	200 910
上　海 Shanghai	5 233 373	16 705 717	490 562	1 513 276
南　通 Nantong	166 198	430 135	4 809	11 083
连云港 Lianyungang	18 929	56 220	533	1 109
杭　州 Hangzhou	697 424	1 794 776	113 424	297 291
宁　波 Ningbo	405 009	749 139	46 647	90 326
温　州 Wenzhou	240 885	532 355	98 766	216 372
福　州 Fuzhou	331 132	1 318 900	72 644	294 026
厦　门 Xiamen	562 657	1 547 984	132 548	412 111
泉　州 Quanzhou	166 741	631 116	346 231	1 089 734
漳　州 Zhangzhou	76 822	169 878	75 095	178 170
青　岛 Qingdao	599 369	2 398 893	122 128	386 284
烟　台 Yantai	279 582	1 252 120	26 007	92 634
威　海 Weihai	280 962	896 817	2 462	5 824
广　州 Guangzhou	3 002 563	7 719 726	3 776 605	8 013 287
深　圳 Shenzhen	1 615 539	3 522 205	9 740 751	19 921 421
珠　海 Zhuhai	523 651	1 133 419	1 125 325	1 931 491
汕　头 Shantou	87 400	229 000	80 200	17 400
湛　江 Zhanjiang	106 100	29 300	96 800	181 900
中　山 Zhongshan	112 330	849 921	355 518	1 276 257
北　海 Beihai	38 227	67 462	24 877	42 430
海　口 Haikou	73 037	129 601	24 663	38 962
三　亚 Sanya	275 955	831 824	59 385	113 074

3-17 续表 continued

城市 City	澳门同胞 Macao		台湾同胞 Taiwan Province	
	人次数 Arrivals	人天数 Nights	人次数 Arrivals	人天数 Nights
天　津 Tianjin	2 867	75 455	44 699	1 121 664
秦皇岛 Qinhuangdao	357	896	5 903	20 302
大　连 Dalian	1 997	7 076	73 703	232 105
上　海 Shanghai	18 952	69 380	653 263	2 443 895
南　通 Nantong	174	368	16 004	54 335
连云港 Lianyungang	21	44	3 489	13 008
杭　州 Hangzhou	15 264	35 350	158 589	382 353
宁　波 Ningbo	6 824	14 058	62 161	122 882
温　州 Wenzhou	55 188	217 841	85 059	182 114
福　州 Fuzhou	6 202	25 119	159 509	348 537
厦　门 Xiamen	7 987	21 081	466 866	1 124 246
泉　州 Quanzhou	33 877	82 197	129 095	298 609
漳　州 Zhangzhou	5 535	10 436	112 884	177 499
青　岛 Qingdao	27 022	75 402	88 086	279 283
烟　台 Yantai	10 054	31 203	43 105	141 576
威　海 Weihai	597	1 432	16 723	53 293
广　州 Guangzhou	486 969	1 029 063	566 853	1 450 765
深　圳 Shenzhen	52 896	97 109	416 730	900 786
珠　海 Zhuhai	721 235	1 618 558	543 156	1 310 797
汕　头 Shantou	700	1 300	8 200	16 700
湛　江 Zhanjiang	7 200	13 700	15 400	42 700
中　山 Zhongshan	78 814	338 900	55 388	380 805
北　海 Beihai	3 704	6 409	7 704	13 653
海　口 Haikou	1 280	1 909	37 930	52 403
三　亚 Sanya	7 137	9 786	46 160	80 756

主要统计指标解释

1. 海洋捕捞产量 凡是从海洋里捕捞的天然生长的水产品产量为捕捞产量。

2. 海水养殖产量 凡是从人工投放苗种或天然纳苗并进行人工饲养管理的海水养殖水域中捕捞的水产品产量为海水养殖产量。

3. 远洋捕捞产量 由各远洋渔业企业和各生产单位按我国远洋渔业项目管理办法组织的远洋渔船（队）在非我国管辖海域（外国专属经济区水域或公海）捕捞的水产品产量。中外合资、合作渔船捕捞的水产品只统计按协议应属于中方所有的部分。

4. 原油产量 是按净原油量来计算的，能直接用于销售和生产自用的原油量。目前海洋石油系统原油产量计算方法采用倒算法。

原油产量=销售量+期末库存量-期初库存量+海上平台及陆地终端处理厂自用量。

5. 天然气产量 指进入集输管网的销售量和就地利用的全部气量。

天然气产量=外输（销）量+企业自用量

6. 造船综合吨 等于以计量单位载重吨和满载排水量吨的民用船舶的吨位数之和。

7. 货运量 指经船舶实际运送的货物重量。

8. 货物周转量 指实际运送的货物与其运送距离的乘积。

9. 集装箱运量 既包含货重，也包含箱重。箱重系指承运租用的空箱重量凡有运费收入的空箱，其重量应统计为运量，按空箱 1 吨为货运量 1 吨计算；若无收入，所承运的空箱一律不作运量统计。

10. 旅客周转量 指实际运送的旅客人数与其运送距离的乘积。

11. 接待人次数 指报告期内我国接待游客人数。游客按出游地分为入境游客和国内游客，按出游时间分为旅游者（过夜游客）和一日游游客（不过夜游客）。

12. 接待人天数 指过夜旅游者的停留天数。

13. 外国人 指外国国籍的人，加入外籍的中国血统华人也计入外国人。

14. 港澳台同胞 指居住在我国香港特别行政区、澳门特别行政区和台湾省的中国同胞。

Explanatory Notes on Main Statistical Indicators

1. Marine Catches refers to the output of the naturally growing aquatic products caught from the sea.

2. Mariculture Production refers to the output of aquatic products whose young are artificially released or naturally collected, and raised and managed artificially, and which are caught from the waters of mariculture.

3. Deep-Sea Fishing Production refers to the output of aquatic products caught in the non-Chinese jurisdictional sea areas (foreign EEE or high sea) by the distant fishing vessels (fleet) organized by various distant fishing businesses and production units according to the management measures of the China distant fishing projects. The aquatic products caught by the Chinese-foreign joint ventures' and cooperative fishing vessels are counted only for the part owned by the Chinese side according to the agreement.

4. Output of Crude Oil is calculated on the basis of the net amount of crude oil, i.e., the amount of crude oil that may be directly used for sale and for the production itself.

Output of crude oil = Volume of sales + Reserves at the end of the period − Reserves at the beginning of the period +Amount for self-use on the platforms and in the terminal processing plants on land.

5. Output of Natural Gas refers to the total gas volume of the sales volume entering the oil collecting and transport pipeline network and that used locally.

Output of natural gas = Volume of sales or transport to other areas + Volume used by the enterprise itself

6. Comprehensive Tonnages of Shipbuilding refers to the sum of tonnage of civilian vessels with the deadweight capacity and full-load displacement as measured.

7. Freight Traffic refers to the weight of cargoes actually transported by vessels.

8. Cargoes Turnover Volume refers to the product of the actually transported cargoes and the transport distance.

9. Freight Volume of Containers includes the weight of both cargoes and container boxes. The weight of container boxes refers to the weight of empty containers rented for transport or having freight income and should be included in the freight volume, one ton of empty boxes equalling to one ton of freight volume. The empty boxes which have no income for transportation are not included in the freight volume.

10. Passenger Turnover Volume refers to the product of the number of passengers actually transported and the shipping distance.

11. Number of Person-Times Received refers to the number of visitors received by China in the period reported. Visitors are divided into inbound visitors and domestic visitors by origin of the travel, and tourists (overnight visitors) and same-day visitors by their length of stay.

12. Number of the Days of Stay refers to the number of the days of stay of tourists.

13. Foreigners refer to the people with foreign nationality, including foreign nationals of Chinese descent.

14. Compatriots from Hong Kong, Macao and Taiwan Province refer to the Chinese compatriots living in the Hong Kong Special Administrative Region, the Macao Special Administrative Region and Taiwan Province.

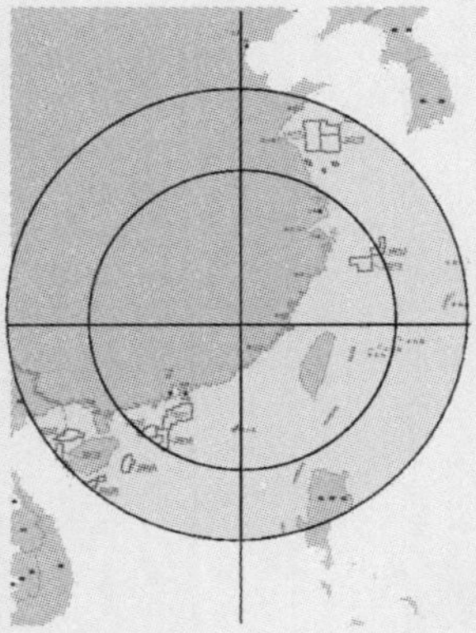

4

主要海洋产业生产能力

Production Capacity of Major Marine Industries

4-1 沿海地区渔港情况
Fishing Ports in the Coastal Regions

单位：个 (unit)

地 区 Region		合计 Total	中心渔港 Central Fishing Port	一级渔港 Grade 1 Fishing Port
合 计	**Total**	**128**	**58**	**70**
天 津	Tianjin			
河 北	Hebei	7	3	4
辽 宁	Liaoning	10	3	7
上 海	Shanghai	1		1
江 苏	Jiangsu	11	6	5
浙 江	Zhejiang	21	9	12
福 建	Fujian	20	8	12
山 东	Shandong	21	11	10
广 东	Guangdong	19	8	11
广 西	Guangxi	8	4	4
海 南	Hainan	10	6	4

资料来源：《2015中国渔业统计年鉴》。
Data Source: *China Fishery Statistical Yearbook 2015*.

4-2 沿海地区海水养殖面积
Mariculture Area by Coastal Regions

单位：公顷 (hm²)

地 区 Region	海水养殖面积 Mariculture Area
合 计 **Total**	**2 305 472**
天 津 Tianjin	3 180
河 北 Hebei	122 434
辽 宁 Liaoning	928 503
上 海 Shanghai	
江 苏 Jiangsu	188 657
浙 江 Zhejiang	88 178
福 建 Fujian	161 418
山 东 Shandong	548 487
广 东 Guangdong	193 691
广 西 Guangxi	54 233
海 南 Hainan	16 691

4-3 海洋油气勘探情况
Work Volume of Offshore Oil and Gas

地 区 Region	地震测线 Seismic Line		钻井（口） Drilling (well)	
	二维 （千米） Two Dimensions (km)	三维 （平方千米） Three Dimensions (km^2)	预探井 Wildcat Wells	评价井 Appraisal Wells
合 计 Total	**21 832**	**15 126**	**72**	**91**
天 津 Tianjin		1 765	17	42
其中：合作 Including: Cooporative			1	
河 北 Hebei	1 018	550	13	11
辽 宁 Liaoning			3	3
上 海 Shanghai	4 098		5	6
其中：合作 Including: Cooporative	3 989		1	
山 东 Shandong	336		3	5
广 东 Guangdong	16 380	12 811	31	24
其中：合作 Including: Cooporative		4 261	2	1

4-4 海洋油气生产井情况
Survey of Offshore Oil and Gas Production Wells

单位：口 (well)

地 区 Region	合 计 Total	采油井 Oil Wells	采气井 Gas Wells	注水井 Injection Wells	其他井 Others
合 计 Total	**6 332**	**4 782**	**330**	**1 209**	**11**
天 津 Tianjin	3 361	2 566	154	641	
河 北 Hebei	1 235	931	13	291	
辽 宁 Liaoning	237	186	6	34	11
上 海 Shanghai	61	18	43		
山 东 Shandong	687	460	8	219	
广 东 Guangdong	751	621	106	24	

4-5 沿海地区盐田面积和海盐生产能力

Salt Pan Area and Sea Salt Production Capacity by Coastal Regions

地 区 Region	盐田总面积（公顷） Total Area of Salt Pan (hm^2)		生产面积（公顷） Production Area (hm^2)		年末海盐生产能力（万吨） Year-End Capacity of Sea Salt Production (10 000 t)	
	2013	2014	2013	2014	2013	2014
合 计 Total	**418 646**	**402 224**	**303 585**	**295 790**	**3 363.79**	**3 446.00**
天 津 Tianjin	27 277	26 923	26 499	26 242	160.00	150.00
河 北 Hebei	77 335	67 992	59 093	58 000	436.35	430.00
辽 宁 Liaoning	33 938	33 664	28 532	28 269	178.67	176.00
江 苏 Jiangsu	54 976	53 200	15 603	11 692	72.00	75.00
浙 江 Zhejiang	2 296	1 971	1 852	1 672	11.66	10.00
福 建 Fujian	5 115	4 116	4 353	4 352	33.00	40.00
山 东 Shandong	202 365	200 890	157 908	157 327	2 431.63	2 540.00
广 东 Guangdong	10 052	8 409	6 014	4 571	16.49	13.00
广 西 Guangxi	1 686	1 600	948	900	5.99	2.00
海 南 Hainan	3 606	3 459	2 783	2 765	18.00	10.00

4-6 2014年海上风电开工项目
Offshore Wind Power Generation Projects in 2014

项 目	开发商	规模（MW）
上海东海大桥二期海上风电场	上海东海风电	116
Second Phase of the Offshore Wind Farm of Donghai Bridge, Shanghai	Shanghai Donghai Wind Power Generations Corporation, Ltd	
江苏滨海海上风电特许招标权项目	大唐	300
Offshore Projects for Wind Power Franchise Bidding Rights in the Coastal Region of Jiangsu	Datang Electric Power Group	
江苏大丰海上风电特许招标权项目	国电龙源	200
Offshore Projects for Wind Power Franchies Bidding Rights, Dafeng, Jiangsu	State Grid Longyuan Electric Power Group Corporation, Ltd	
江苏东台海上风电特许招标权项目	鲁能集团	200
Offshore Projects for Wind Power Franchise Bidding Rights, Dongtai, Jiangsu	Luneng Group Corporation, Ltd	
江苏如东潮间带风电场	国电龙源	200
Tidal Zone Wind Farm, Rudong, Jiangsu	State Grid Longyuan Electric Power Group Corporation, Ltd	
江苏如东海上风电场	中广核	150
Offshore Wind Farm, Rudong, Shanghai	Zhongguanghe Wind Power Generation Corporation, Ltd	
江苏响水海上风电场	三峡新能源	200
Offshore Wind Farm, Xiangshui, Jiangsu	Three Gorges New Energy Corporation	
广东珠海桂山海上风电场	南方电网	200
Guishan Offshore Wind Farm, Zhuhai, Guangdong	China Southern Grid	
福建莆田平海湾海上风电场	福建中闽	300
Offshore Wind Farm, Putian Gulf, Fujian	Fujian Zhongmin Energy Investment Corporation, Ltd	

4-7 主要潮汐电站分布情况
Distribution of Major Tidal Power Stations

电站名称 Name	运行情况 Status of Operation	装机容量（千瓦） Capacity (kW)
江厦潮汐试验电站	1972年开始建造，1980年投入使用，运行至今	3 900
Jiangxia Experimental Tidal Power Station	It began construction in 1972, was put into use in 1980, and has been in operation so far	
海山潮汐电站	1972年开始建造，1975年投入使用，运行至今	250
Haishan Tidal Power Station	It began construction in 1972, was put into use in 1975, and has been in operation so far	
岳浦潮汐电站	1970年开始建造，1978年停止运行	150
Yuepu Tidal Power Station	It began construction in 1970, and stopped power generation in 1978	
白沙口潮汐电站	1970年开始建造，1978年投入使用，2010年停止运行	960
Baishakou Tidal Power Station	It began construction in 1970, was put into use in 1978, and stopped power generation in 2010	

4-8 主要海上活动船舶
Major Vessels Operating on the Sea

类 别 Type	艘数 （艘） Number of Vessels (unit)	总吨 （万吨位） Gross Tonnage (10 000 t)	净载重量 （万吨） Net Weight Tonnage (10 000 t)	载客量 （客位） Passenger Spaces (seat)	总功率 （千瓦） Total Power (kW)
一 、海洋生产用船 Vessels for Marine Production					
海洋渔业船舶 Marine Fishing Vessels	277 453	853.9			16 962 042
远洋渔船 Ocean-going Fishing Vessels	2 460				2 025 562
海洋油气船舶 Offshore Oil and Gas Vessels	175	112.0	53.0	10 753	997 354
钻井平台 Drilling Vessels	45	46.0	14.0	6 060	333 047
物探船 Physical Exploration Vessels	9	4.0	1.0	495	51 177
其 他 Others	121	62.0	38.0	4 198	613 130
海洋运输船舶 Maritime Transport Vessels	12 916		14 109.2	224 580	38 119 681
二、海洋科研用船 Vessels for Marine Scientific Research					
海洋地质勘探船 Marine Geological Survey Vessels	6	1.2	0.4	425	23 371
海洋调查船 Marine Research Vessels					
中国科学院 Chinese Acadamy of Sciences	7	4.3	1.3	238	30 504
国家海洋局 State Oceanic Administration	9	4.6		389	37 749

4-9 沿海规模以上港口生产用码头泊位

Berths for Productive Use at the coastal Seaports above Designed Size

单位：米, 个 (m, unit)

港口 Seaport		码头长度 Length of Quay Line	泊位个数 Number of Berths	#万吨级 10 000 Tonnage Class
合 计	**Total**	**701 232**	**4 970**	**1 614**
丹 东	Dandong	7 378	41	24
大 连	Dalian	38 730	215	98
营 口	Yingkou	17 432	80	54
锦 州	Jinzhou	6 119	23	21
秦皇岛	Qinhuangdao	15 928	72	44
黄 骅	Huanghua	8 761	36	30
唐 山	Tangshan	23 715	89	86
天 津	Tianjin	34 606	151	106
烟 台	Yantai	17 911	88	59
威 海	Weihai	4 455	22	14
青 岛	Qingdao	21 696	84	68
日 照	Rizhao	13 198	51	45
上 海	Shanghai	75 091	608	156
连云港	Lianyungang	14 356	60	53
嘉 兴	Jiaxing	10 458	80	32
宁波-舟山	Ningbo-Zhoushan	80 125	615	150
台 州	Taizhou	12 309	180	8
温 州	Wenzhou	17 323	218	18

4-9 续表 continued

港 口 Seaport		码头长度 Length of Quay Line	泊位个数 Number of Berths	#万吨级 10 000 Tonnage Class
福 州	Fuzhou	22 726	168	52
莆 田	Putian	5 002	46	8
泉 州	Quanzhou	15 991	104	25
厦 门	Xiamen	27 216	154	69
汕 头	Shantou	9 627	87	19
汕 尾	Shanwei	1 286	14	1
惠 州	Huizhou	9 048	39	18
深 圳	Shenzhen	28 825	141	66
虎 门	Humen	14 665	112	26
广 州	Guangzhou	46 958	510	69
中 山	Zhongshan	3 332	57	0
珠 海	Zhuhai	17 768	155	27
江 门	Jiangmen	10 193	152	3
阳 江	Yangjiang	2 232	10	9
茂 名	Maoming	2 428	18	9
湛 江	Zhanjiang	15 542	146	30
北部湾港	Beibuwan	33 509	249	74
海 口	Haikou	6 858	51	15
洋 浦	Yangpu	6 281	32	19
八 所	Basuo	2 154	12	9

4-10 沿海地区星级饭店基本情况
Star Grade Hotels and Occupancies by Coastal Regions

地　区 Region		饭店数（座） Number of Hotels (unit)	客房数（间） Number of Rooms (unit)	床位数（张） Number of Beds (unit)	客房出租率 （%） Room Occupancy (%)
合　计	**Total**	**5 035**	**749 993**	**1 235 293**	
天　津	Tianjin	93	18 527	28 867	49.21
河　北	Hebei	391	52 123	92 756	44.47
辽　宁	Liaoning	405	52 426	88 194	48.65
上　海	Shanghai	240	57 428	86 777	63.70
江　苏	Jiangsu	650	89 442	144 316	57.85
浙　江	Zhejiang	792	114 016	190 586	54.76
福　建	Fujian	374	59 182	94 266	57.82
山　东	Shandong	724	91 089	156 454	55.88
广　东	Guangdong	832	136 580	215 023	55.41
广　西	Guangxi	401	53 540	93 734	53.99
海　南	Hainan	133	25 640	44 320	56.37

注：本表数据为主要沿海城市合计数。
Note: The data in this table are for the total number of major coastal cities.

4-11　沿海地区旅行社数
Number of Travel Agencies by Coastal Regions

单位：家　　(unit)

地　区 Region		旅行社总数 Number of Travel Agencies
合　计	**Total**	**13 790**
天　津	Tianjin	377
河　北	Hebei	1 343
辽　宁	Liaoning	1 210
上　海	Shanghai	1 185
江　苏	Jiangsu	2 099
浙　江	Zhejiang	2 036
福　建	Fujian	805
山　东	Shandong	2 054
广　东	Guangdong	1 792
广　西	Guangxi	537
海　南	Hainan	352

主要统计指标解释

1. 海水养殖面积 是指利用海上、滩涂、陆基进行鱼、甲壳类（虾、蟹）、贝、藻等海水经济动植物的人工养殖的水面面积。在报告期内无论是否全部收获或尚未收获其产品，均应统计在海水养殖面积中。但有些滩涂、水面不投放苗种或投放少量苗种，只进行一般管理的，不统计为养殖面积。

2. 盐田总面积 指盐田占有的全部面积。包括储卤、蒸发、保卤、结晶面积、滩内的沟、壕、池、埝、滩坨等面积及滩外的沟、壕、公路及杂地面积。

3. 生产面积 指直接提供给海盐生产的面积，包括结晶面积、蒸发面积、保卤面积，滩内的沟、壕、池、埝面积及滩坨面积。

4. 年末海盐生产能力 指年末企业生产原盐的全部设备的综合平衡能力。海盐生产露天作业，受天气影响，因而计算生产能力时，成熟滩田按 10 年实际平均单位生产面积产量乘以本年成熟滩田生产面积而得，新滩田按设计能力及滩田成熟程度可能达到的产量计算。

5. 海洋渔业船舶 是指配置机器作为动力的从事海洋渔业生产和辅助渔业生产的船舶。

6. 远洋渔船 按我国远洋渔业项目管理办法在非我国管辖海域（外国专属经济区水域或公海）进行常年或季节性生产的渔船。

7. 泊位个数 是指设有系靠船舶设施，在同一时间内可供靠泊最大吨级船舶的艘数。即可靠泊一艘船舶，则计为一个泊位，余类推。泊位分码头泊位和浮筒泊位。

8. 客房数 指饭店实际可用于接待旅游者的房间数。

9. 床位数 指饭店实际可用于接待旅游者的床位数。

Explanatory Notes on Main Statistical Indicators

1. Mariculture Area refers to the area of the water surface where seawater economic animals and plants, such as crustacean (shrimp, crab), shellfish and algae, are cultivated at sea, on tidal flat and land. Whether or not all the products in the area have been harvested or the products have not been harvested yet in the period covered by the report, the area is included in the Mariculture Area. But some tidal flats and water surfaces where none or a small amount of the young have been released and only general management is carried out are not included in the Mariculture Area.

2. Total Area of Salt Pans refers to the total area covered by salt pans, including the area for brine storage, evaporation, brine preservation, and crystallization, the area of ditches, moats, ponds and banks within the beach as well as beach mounds, and ditches, moats, highway beyond the beach as well as the area of miscellaneous lands.

3. Area of Salt Pan Production refers to the area directly provided for sea-salt productions, including the area for crystallization, evaporation and brine preservation, the area of ditches, moats, ponds and banks within the beach as well as the area of beach mounds.

4. Year-End Capacity of Crude Salt Production refers to the integrated and balanced capacity of all equipment of the enterprise used for crude salt production at the end of the year. As sea salt production is an open-air operation, which is subject to the effect of weather, the production capacity of a matured salt pan is calculated at the productions of the actual average unit production area in ten years times the production area of the matured salt pan in the current year. The production capacity of new salt pans is calculated at the production that may be reached in the light of the designed capacity and the level of maturity of the salt pan.

5. Marine Fishing Vessels refer to the vessels equipped with machines as motive power and going for marine fishery production and auxiliary fishery production.

6. Deep Sea Fishing Vessels refer to the fishing vessels which carry out production all the year round or seasonally in the non-Chinese jurisdictional, sea areas (foreign EEZ or high sea) according to the China Deep-Sea Fishing Projects Management Measures.

7. Number of Berths refers to the spaces equipped with facilities for docking ships and the number of ships of the maximum tonnage that may dock or anchor in them. A space for a ship to dock is counted as one berth and the rest are reasoned out by analogy. Berths are divided into wharf berths and buoy berths.

8. Number of Rooms refers to the number of guest rooms actually used by the hotels receiving tourists.

9. Number of Beds refers to the number of beds actually used by the hotels receiving tourists.

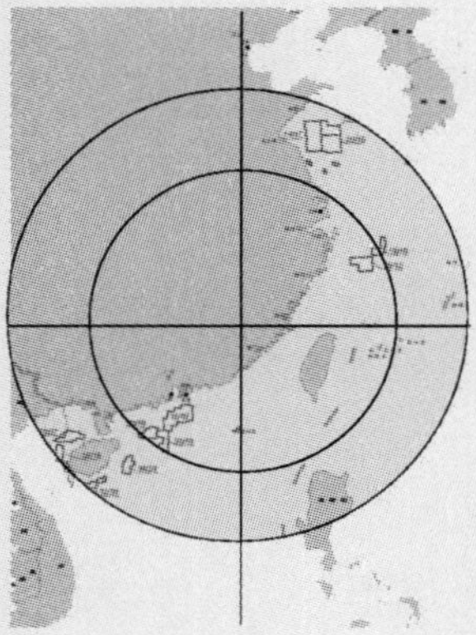

5

涉 海 就 业

Ocean-Related Employment

5-1 全国涉海就业人员情况
National Ocean-Related Employed Personnel by Region

单位：万人 (10 000 persons)

地 区 Region	2001	2013①	2014①
合 计 Total	**2 107.6**	**3 514.3**	**3 553.7**
天 津 Tianjin	106.4	177.4	179.4
河 北 Hebei	58.0	96.7	97.8
辽 宁 Liaoning	196.0	326.8	330.5
上 海 Shanghai	127.5	212.6	215.0
江 苏 Jiangsu	116.9	194.9	197.1
浙 江 Zhejiang	256.4	427.5	432.3
福 建 Fujian	259.7	433.0	437.9
山 东 Shandong	319.9	533.4	539.4
广 东 Guangdong	505.3	842.6	852.0
广 西 Guangxi	68.9	114.9	116.2
海 南 Hainan	80.6	134.4	135.9
其 他② Others	12.0	20.0	20.2

注：① 2013年和2014年为推算数据；② 其他为非沿海地区涉海就业人员数。

Note: ① The data for 2013 and 2014 are the estimated ones;

② Others refer to the number of ocean-related employed persons in the non-coastal regions.

5-2 全国主要海洋产业就业人员情况
Employed Personnel in the Major Marine Industries Throughout the Country

单位：万人 (10 000 persons)

海洋产业 Marine Industry	2001	2013	2014
合计 Total	**719.1**	**1 199.1**	**1212.5**
海洋渔业及相关产业 Marine Fishery and the Related Industries	348.3	580.8	587.3
海洋石油和天然气业 Offshore Oil and Natural Gas Industry	12.4	20.7	20.9
海滨砂矿业 Beach Placer Mining Industry	1.0	1.7	1.7
海洋盐业 Sea Salt Industry	15.0	25.0	25.3
海洋化工业 Marine Chemical Industry	16.1	26.8	27.1
海洋生物医药业 Marine Biomedicine Industry	0.6	1.0	1.0
海洋电力和海水利用业 Marine Electric Power and Seawater Utilization Industry	0.7	1.2	1.2
海洋船舶工业 Marine Shipbuilding Industry	20.6	34.3	34.7
海洋工程建筑业 Marine Engineering Architecture Industry	38.8	64.7	65.4
海洋交通运输业 Maritime Communications and Transportation Industry	50.8	84.7	85.7
滨海旅游业 Coastal Tourism	78.3	130.6	132.0
其他海洋产业 Other Marine Industries	136.5	227.6	230.2

注：2013年和2014年为推算数据。

Note: The data for 2013 and 2014 are the estimated ones.

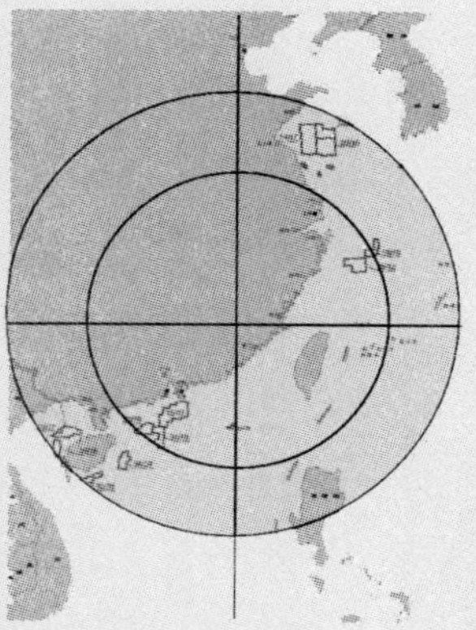

6

海洋科学技术

Marine Science and Technology

6-1 分行业海洋科研机构及人员情况
Marine Scientific Research Institutions and Personnel by Industry

行　业 Industry	机构数（个） Number of Institutions (unit)	从业人员（人） Employed Population (person)
合 计 Total	**189**	**40 539**
海洋基础科学研究 Marine Basic Scientific Research	**104**	**18 441**
海洋自然科学 Marine Natural Science	62	14 121
海洋社会科学 Marine Social Science	3	781
海洋农业科学 Marine Agricultural Science	37	3 446
海洋生物医药 Marine Biomedicine	2	93
海洋工程技术研究 Marine Engineering Technology Research	**73**	**20 375**
海洋化学工程技术 Marine Chemical Engineering Technology	12	6 322
海洋生物工程技术 Marine Bioengineering Technology	2	230
海洋交通运输工程技术 Marine Communications and Transport Technology	19	5 361

6-1 续表 continued

行 业 Industry	机构数（个） Number of Institutions (unit)	从业人员（人） Employed Population (person)
海洋能源开发技术 Marine Energies Development Technology	4	3 150
海洋环境工程技术 Marine Environmental Engineering Technology	14	1 194
河口水利工程技术 Estuarine Water Conservancy Engineering Technology	18	3 319
其他海洋工程技术 Other Marine Engineering Technology	4	799
海洋信息服务业 Marine Information Service	**9**	**1 050**
其他海洋信息服务 Other Marine Information Services	9	1 050
海洋技术服务业 Marine Technological Service Industry	**3**	**673**
海洋工程管理服务 Marine Engineering Management Service	1	543
其他海洋专业技术服务 Other Marine Professional and Technological Services	2	130

注：机构为县级以上科研机构；行业分类参照《海洋及相关产业分类》标准。

Note: The institutions are the scientific research institutions above the county level;The classification of industries follows the standard *Classification of Marine Industries and the Related Industries*.

6-2 分行业海洋科研机构科技活动人员学历构成

Educational Background Composition of the Personnel Engaged in Scientific and Technological Activities in Marine Scientific Research Institutions by Industry

单位：人 (person)

行　业 Industry	科技活动人员 Personnel Engaged in Scientific Activities	博士 Doctor	硕士 Master	大学生 University Graduate	大专生 College Graduate
合　计 Total	**34 174**	**8 277**	**10 386**	**10 069**	**2 931**
海洋基础科学研究 Marine Basic Scientific Research	**16 766**	**5 535**	**4 649**	**4 169**	**1 251**
海洋自然科学 Marine Natural Science	13 132	4 943	3 521	2 841	892
海洋社会科学 Marine Social Science	742	130	298	224	49
海洋农业科学 Marine Agricultural Science	2 819	461	820	1 054	301
海洋生物医药 Marine Biomedicine	73	1	10	50	9
海洋工程技术研究 Marine Engineering Technology Research	**15 743**	**2 613**	**5 115**	**5 283**	**1 531**
海洋化学工程技术 Marine Chemical Engineering Technology	5 359	811	1 327	1 647	683
海洋生物工程技术 Marine Bioengineering Technology	205	18	94	68	21

6-2 续表 continued

行业 Industry	科技活动人员 Personnel Engaged in Scientific Activities	博士 Doctor	硕士 Master	大学生 University Graduate	大专生 College Graduate
海洋交通运输工程技术 Marine Communications and Transport Technology	3 192	189	1 214	1 322	330
海洋能源开发技术 Marine Energies Development Technology	2 617	911	905	551	191
海洋环境工程技术 Marine Environmental Engineering Technology	1 057	57	337	571	77
河口水利工程技术 Estuarine Water Conservancy Engineering Technology	2 741	541	925	977	209
其他海洋工程技术 Other Marine Engineering Technology	572	86	313	147	20
海洋信息服务业 Marine Information Service	**1 011**	**86**	**371**	**322**	**125**
其他海洋信息服务 Other Marine Information Services	1 011	86	371	322	125
海洋技术服务业 Marine Technological Service Industry	**654**	**43**	**251**	**295**	**24**
海洋工程管理服务 Marine Engineering Management Service	543	38	236	250	17
其他海洋专业技术服务 Other Marine Professional and Technological Services	111	5	15	45	7

6-3 分行业海洋科研机构科技活动人员职称构成

Professional Title Composition of Personnel Engaged in Scientific and Technological Activities in the Marine Scientific Research Institutions by Industry

单位：人 (person)

行 业 Industry	科技活动人员 Personnel Engaged in Scientific Activities	高级职称 Senior	中级职称 Intermediate	初级职称 Primary
合 计 Total	**34 174**	**14 161**	**11 692**	**4 891**
海洋基础科学研究 Marine Basic Scientific Research	**16 766**	**7 033**	**6 061**	**2 377**
海洋自然科学 Marine Natural Science	13 132	5 659	4 859	1 690
海洋社会科学 Marine Social Science	742	347	185	62
海洋农业科学 Marine Agricultural Science	2 819	1 017	998	584
海洋生物医药 Marine Biomedicine	73	10	19	41
海洋工程技术研究 Marine Engineering Technology Research	**15 743**	**6 544**	**5 034**	**2 233**
海洋化学工程技术 Marine Chemical Engineering Technology	5 359	2 110	1 653	597
海洋生物工程技术 Marine Bioengineering Technology	205	61	83	60

6-3 续表 continued

行业 Industry	科技活动人员 Personnel Engaged in Scientific Activities	高级职称 Senior	中级职称 Intermediate	初级职称 Primary
海洋交通运输工程技术 Marine Communications and Transport Technology	3 192	984	1 071	786
海洋能源开发技术 Marine Energies Development Technology	2 617	1 435	794	205
海洋环境工程技术 Marine Environmental Engineering Technology	1 057	392	403	205
河口水利工程技术 Estuarine Water Conservancy Engineering Technology	2 741	1 331	792	308
其他海洋工程技术 Other Marine Engineering Technology	572	231	238	72
海洋信息服务业 **Marine Information Service**	**1 011**	**332**	**309**	**220**
其他海洋信息服务 Other Marine Information Services	1 011	332	309	220
海洋技术服务业 **Marine Technological Service Industry**	**654**	**252**	**288**	**61**
海洋工程管理服务 Marine Engineering Management Service	543	214	261	54
其他海洋专业技术服务 Other Marine Professional and Technological Services	111	38	27	7

6-4 分行业海洋科研机构经费收入

Research Funds Receipts of the Marine Scientific Research Institutions by Industry

单位：千元 (1 000 yuan)

行 业 Industry	经费收入总额 Fund Total	经常费 Routine Fund	科技活动借贷款 Loans in the Scientific and Technological Activities	基本建设中政府投资 Government Investment in the Capital Construction
合 计 **Total**	**31 009 893**	**29 203 821**	**85 825**	**1 720 247**
海洋基础科学研究 **Marine Basic Scientific Research**	**14 756 965**	**13 413 772**	**5 075**	**1 338 118**
海洋自然科学 Marine Natural Science	12 800 787	11 553 395	5 075	1 242 317
海洋社会科学 Marine Social Science	420 126	401 812	0	18 314
海洋农业科学 Marine Agricultural Science	1 513 383	1 435 896	0	77 487
海洋生物医药 Marine Biomedicine	22 669	22 669	0	0
海洋工程技术研究 **Marine Technological Service Industry**	**15 254 272**	**14 795 905**	**80 750**	**377 617**
海洋化学工程技术 Marine Chemical Engineering Technology	4 280 639	4 193 687	0	86 952
海洋生物工程技术 Marine Bioengineering Technology	183 575	183 575	0	0

6-4 续表 continued

行 业 Industry	经费收入总额 Fund Total	经常费 Routine Fund	科技活动借贷款 Loans in the Scientific and Technological Activities	基本建设中政府投资 Government Investment in the Capital Construction
海洋交通运输工程技术 Marine Communications and Transport Technology	4 693 687	4 413 075	80 000	200 612
海洋能源开发技术 Marine Energies Development Technology	2 832 846	2 832 846	0	0
海洋环境工程技术 Marine Environmental Engineering Technology	732 313	684 666	750	46 897
河口水利工程技术 Estuarine Water Conservancy Engineering Technology	1 885 291	1 842 135	0	43 156
其他海洋工程技术 Other Marine Engineering Technology	645 921	645 921	0	0
海洋信息服务业 Marine Information Service	**642 891**	**638 379**	**0**	**4 512**
其他海洋信息服务 Other Marine Information Services	642 891	638 379	0	4 512
海洋技术服务业 Marine Technological Service Industry	**355 765**	**355 765**	**0**	**0**
海洋工程管理服务 Marine Engineering Management Service	101 360	101 360	0	0
其他海洋专业技术服务 Other Marine Professional and Technological Services	254 405	254 405	0	0

6-5 分行业海洋科研机构科技课题情况
Marine Scientific and Technological Research Projects of the Research Institutions by Industry

单位：项 (item)

行　业 Industry	课题数 Number of Research Projects	基础研究 Basic Research	应用研究 Applied Research	试验发展 Experimental Development	成果应用 Result Application	科技服务 Scientific and Technological Service
合　计 Total	**17 702**	**4 447**	**4 122**	**4 480**	**1 861**	**2 792**
海洋基础科学研究 Marine Basic Scientific Research	**13 140**	**4 331**	**3 618**	**2 535**	**1 192**	**1 464**
海洋自然科学 Marine Natural Science	10 707	4 128	3 360	1 865	444	910
海洋社会科学 Marine Social Science	827	8	18	108	395	298
海洋农业科学 Marine Agricultural Science	1 589	194	238	557	344	256
海洋生物医药 Marine Biomedicine	17	1	2	5	9	0
海洋工程技术研究 Marine Engineering Technology Research	**4 341**	**116**	**503**	**1 889**	**665**	**1 168**
海洋化学工程技术 Marine Chemical Engineering Technology	1 043	28	150	656	170	39
海洋生物工程技术 Marine Bioengineering Technology	75	0	2	20	44	9

6-5 续表 continued

行 业 Industry	课题数 Number of Research Projects	基础研究 Basic Research	应用研究 Applied Research	试验发展 Experimental Development	成果应用 Result Application	科技服务 Scientific and Technological Service
海洋交通运输工程技术 Marine Communications and Transport Technology	941	9	47	384	151	350
海洋能源开发技术 Marine Energies Development Technology	685	19	105	339	39	183
海洋环境工程技术 Marine Environmental Engineering Technology	274	4	30	76	43	121
河口水利工程技术 Estuarine Water Conservancy Engineering	1 167	54	167	318	181	447
其他海洋工程技术 Other Marine Engineering Technology	156	2	2	96	37	19
海洋信息服务业 Marine Information Service	**156**	**0**	**0**	**38**	**1**	**117**
其他海洋信息服务 Other Marine Information Services	156	0	0	38	1	117
海洋技术服务业 Marine Technological Service Industry	**65**	**0**	**1**	**18**	**3**	**43**
海洋工程管理服务 Marine Engineering Management Service	50	0	0	7	0	43
其他海洋专业技术服务 Other Marine Professional and Technological Services	15	0	1	11	3	0

6-6 分行业海洋科研机构科技论著情况
Marine Scientific and Technological Works of the Research Institutions by Industry

行 业 Industry	发表科技论文（篇） Scientific Theses Published (piece)	#国外发表 Published Abroad	出版科技著作（种） Scientific and Technological Works Published (kind)
合 计 Total	**16 908**	**5 848**	**314**
海洋基础科学研究 Marine Basic Scientific Research	**11 590**	**4 997**	**195**
海洋自然科学 Marine Natural Science	9 015	4 547	112
海洋社会科学 Marine Social Science	841	20	55
海洋农业科学 Marine Agricultural Science	1 732	430	28
海洋生物医药 Marine Biomedicine	2	0	0
海洋工程技术研究 Marine Engineering Technology Research	**4 882**	**833**	**112**
海洋化学工程技术 Marine Chemical Engineering Technology	897	156	5
海洋生物工程技术 Marine Bioengineering Technology	79	5	1

6-6 续表 continued

行 业 Industry	发表科技论文（篇） Scientific Theses Published (piece)	#国外发表 Published Abroad	出版科技著作（种） Scientific and Technological Works Published (kind)
海洋交通运输工程技术 Marine Communications and Transport Technology	1 054	187	12
海洋能源开发技术 Marine Energies Development Technology	1 205	212	26
海洋环境工程技术 Marine Environmental Engineering Technology	199	22	5
河口水利工程技术 Estuarine Water Conservancy Engineering	1 032	186	52
其他海洋工程技术 Other Marine Engineering Technology	416	65	11
海洋信息服务业 **Marine Information Service**	**329**	**18**	**5**
其他海洋信息服务 Other Marine Information Services	329	18	5
海洋技术服务业 **Marine Technological Service Industry**	**107**	**0**	**2**
海洋工程管理服务 Marine Engineering Management Service	80	0	0
其他海洋专业技术服务 Other Marine Professional and Technological Services	27	0	2

6-7 分行业科研机构科技专利情况
Marine Scientific and Technological Patents of the Research Institutions by Industry

单位：件 (unit)

行 业 Industry	专利申请受理数 Number of Patent Applications Accepted	#发明专利 Patents for Discoveries	专利授权数 Number of Patents Granted	#发明专利 Patents for Discoveries	拥有发明专利总数 Total Number of Patents for Discoveries
合 计 Total	**6 111**	**4 813**	**4 020**	**2 647**	**13 966**
海洋基础科学研究 Marine Basic Scientific Research	**2 383**	**1 643**	**1 576**	**810**	**3 788**
海洋自然科学 Marine Natural Science	1 694	1 283	1 032	628	3 079
海洋社会科学 Marine Social Science	0	0	9	1	22
海洋农业科学 Marine Agricultural Science	689	360	535	181	687
海洋生物医药 Marine Biomedicine	0	0	0	0	0
海洋工程技术研究 Marine Engineering Technology Research	**3 701**	**3 154**	**2 438**	**1 832**	**10 132**
海洋化学工程技术 Marine Chemical Engineering Technology	2 665	2 532	1 856	1 643	9 106
海洋生物工程技术 Marine Bioengineering Technology	1	1	0	0	3

6-7 续表 continued

行 业 Industry	专利申请受理数 Number of Patent Applications Accepted	#发明专利 Patents for Discoveries	专利授权数 Number of Patents Granted	#发明专利 Patents for Discoveries	拥有发明专利总数 Total Number of Patents for Discoveries
海洋交通运输工程技术 Marine Communications and Transport Technology	205	91	143	32	182
海洋能源开发技术 Marine Energies Development Technology	398	308	168	87	391
海洋环境工程技术 Marine Environmental Engineering Technology	28	10	13	4	21
河口水利工程技术 Estuarine Water Conservancy Engineering Technology	190	80	164	35	161
其他海洋工程技术 Other Marine Engineering Technology	214	132	94	31	268
海洋信息服务业 Marine Information Service	**7**	**6**	**1**	**1**	**3**
其他海洋信息服务 Other Marine Information Services	7	6	1	1	3
海洋技术服务业 Marine Technological Service Industry	**20**	**10**	**5**	**4**	**43**
海洋工程管理服务 Marine Engineering Management Service	2	1	2	2	5
其他海洋专业技术服务 Other Marine Professional and Technological Services	18	9	3	2	38

6-8 分行业海洋科研机构R & D情况
R & D in the Marine Scientific Research Institutions by Profession

行　业 Industry	R & D人员 （人） R & D Personnel (person)	R & D经费内部支出 （千元） R & D Internal Expenditure (1 000 yuan)	R & D课题数 （项） Number of R & D Projects (item)
合　计 Total	**28 243**	**15 663 775**	**13 049**
海洋基础科学研究 Marine Basic Scientific Research	**18 414**	**9 643 753**	**10 484**
海洋自然科学 Marine Natural Science	16 219	8 841 972	9 353
海洋社会科学 Marine Social Science	371	83 015	134
海洋农业科学 Marine Agricultural Science	1 803	715 791	989
海洋生物医药 Marine Biomedicine	21	2 975	8
海洋工程技术研究 Marine Engineering Technology Research	**9 211**	**5 769 548**	**2 508**
海洋化学工程技术 Marine Chemical Engineering Technology	3 220	2 217 948	834
海洋生物工程技术 Marine Bioengineering Technology	39	21 892	22

6-8 续表 continued

行　业 Industry	R & D人员 （人） R & D Personnel (person)	R & D经费内部支出 （千元） R & D Internal Expenditure (1 000 yuan)	R & D课题数 （项） Number of R & D Projects (item)
海洋交通运输工程技术 Marine Communications and Transport Technology	1 968	844 026	440
海洋能源开发技术 Marine Energies Development Technology	1 827	1 690 421	463
海洋环境工程技术 Marine Environmental Engineering Technology	399	179 585	110
河口水利工程技术 Estuarine Water Conservancy Engineering Technology	1 369	641 552	539
其他海洋工程技术 Other Marine Engineering Technology	389	174 124	100
海洋信息服务业 Marine Information Service	**467**	**173 593**	**38**
其他海洋信息服务 Other Marine Information Services	467	173 593	38
海洋技术服务业 Marine Technological Service Industry	**151**	**76 881**	**19**
海洋工程管理服务 Marine Engineering Management Service	50	16 175	7
其他海洋专业技术服务 Other Marine Professional and Technological Services	101	60 706	12

6-9　分地区海洋科研机构及人员情况
Marine Scientific Research Institutions and Personnel by Region

地　区 Region	机构数（个） Number of Institutions (unit)	从业人员（人） Employed Population (person)
合 计 **Total**	**189**	**40 539**
北 京 Beijing	24	14 091
天 津 Tianjin	14	2 772
河 北 Hebei	5	547
辽 宁 Liaoning	22	2 246
上 海 Shanghai	15	3 866
江 苏 Jiangsu	11	3 161
浙 江 Zhejiang	20	1 914
福 建 Fujian	14	1 156
山 东 Shandong	21	3 922
广 东 Guangdong	25	3 835
广 西 Guangxi	11	1 199
海 南 Hainan	3	277
其 他 Others	4	1 553

6-10 分地区海洋科研机构科技活动人员学历构成

Educational Background Composition of the Personnel Engaged in Scientific and Technological Activities in Marine Scientific Research Institutions by Region

单位：人 (person)

地 区 Region	科技活动人员 Personnel Engaged in Scientific Activities	博士 Doctor	硕士 Master	大学生 University Graduate	大专生 College Graduate
合 计 Total	**34 174**	**8 277**	**10 386**	**10 069**	**2 931**
北 京 Beijing	12 603	4 055	3 746	2 881	889
天 津 Tianjin	2 269	203	791	932	181
河 北 Hebei	515	49	139	236	49
辽 宁 Liaoning	1 836	188	515	734	240
上 海 Shanghai	3 484	616	1 125	1 157	324
江 苏 Jiangsu	1 954	383	530	632	230
浙 江 Zhejiang	1 638	180	592	674	124
福 建 Fujian	1 100	159	329	391	136
山 东 Shandong	3 338	877	961	935	367
广 东 Guangdong	3 292	920	1 042	857	249
广 西 Guangxi	627	37	200	308	56
海 南 Hainan	240	11	73	118	32
其 他 Others	1 278	599	343	214	54

6-11 分地区海洋科研机构科技活动人员职称构成
Professional Title Composition of Personel Engaged in Marine Scientific Research Institutions by Region

单位：人 (person)

地 区 Region	科技活动人员 Personnel Engaged in Scientific Activities	高级职称 Senior	中级职称 Intermediate	初级职称 Primary
合 计 Total	**34 174**	**14 161**	**11 692**	**4 891**
北 京 Beijing	12 603	6 054	4 107	1 218
天 津 Tianjin	2 269	889	744	417
河 北 Hebei	515	213	118	29
辽 宁 Liaoning	1 836	698	597	214
上 海 Shanghai	3 484	1 264	1 347	649
江 苏 Jiangsu	1 954	852	522	376
浙 江 Zhejiang	1 638	599	527	264
福 建 Fujian	1 100	365	390	222
山 东 Shandong	3 338	1 218	1 324	592
广 东 Guangdong	3 292	1 233	1 169	501
广 西 Guangxi	627	190	255	147
海 南 Hainan	240	56	51	100
其 他 Others	1 278	530	541	162

6-12 分地区海洋科研机构经费收入
Research Funds Receipts of Marine Scientific Research Institutions by Region

单位：千元 (1 000 yuan)

地　区 Region	经费收入总额 Fund Total	经常费 Routine Fund	科技活动借贷款 Loans in the Scientific and Technological Activities	基本建设中政府投资 Government Investment in the Capital Construction
合 计 Total	**31 009 893**	**29 203 821**	**85 825**	**1 720 247**
北 京 Beijing	10 801 649	10 515 018	5 075	281 556
天 津 Tianjin	1 677 311	1 530 416	0	146 895
河 北 Hebei	142 818	139 865	0	2 953
辽 宁 Liaoning	1 296 498	1 279 539	0	16 959
上 海 Shanghai	3 638 014	3 382 583	0	255 431
江 苏 Jiangsu	2 440 990	2 303 165	80 000	57 825
浙 江 Zhejiang	1 449 886	1 354 587	750	94 549
福 建 Fujian	1 178 354	1 042 731	0	135 623
山 东 Shandong	3 818 248	3 227 713	0	590 535
广 东 Guangdong	2 752 209	2 630 811	0	121 398
广 西 Guangxi	833 255	823 782	0	9 473
海 南 Hainan	99 138	98 888	0	250
其 他 Others	881 523	874 723	0	6 800

6-13 分地区海洋科研机构科技课题情况
Marine Scientific and Technological Research Projects of the Research Institutions by Region

单位：项 (item)

地区 Region	课题数 Number of Research Projects	基础研究 Basic Research	应用研究 Applied Research	试验发展 Experimental Development	成果应用 Result Application	科技服务 Scientific and Technological Service
合计 Total	**17 702**	**4 447**	**4 122**	**4 480**	**1 861**	**2 792**
北京 Beijing	6 655	1 865	1 517	1 248	670	1 355
天津 Tianjin	751	17	51	346	87	250
河北 Hebei	90	9	23	31	18	9
辽宁 Liaoning	391	4	72	183	96	36
上海 Shanghai	1 131	112	288	329	200	202
江苏 Jiangsu	2 202	66	597	910	335	294
浙江 Zhejiang	619	115	55	110	127	212
福建 Fujian	608	184	217	117	55	35
山东 Shandong	1 633	393	563	360	184	133
广东 Guangdong	2 140	687	624	554	43	232
广西 Guangxi	101	10	17	66	7	1
海南 Hainan	10	0	0	0	10	0
其他 Others	1 371	985	98	226	29	33

6-14 分地区海洋科研机构科技论著情况
Marine Scientific and Technological Works of the Research Institutions by Region

地 区 Region	发表科技论文（篇） Scientific Theses Published (piece)	#国外发表 Published Abroad	出版科技著作（种） Scientific and Technological Works Published (kind)
合 计 Total	**16 908**	**5 848**	**314**
北 京 Beijing	5 909	2 111	115
天 津 Tianjin	1 038	193	14
河 北 Hebei	494	5	48
辽 宁 Liaoning	442	37	14
上 海 Shanghai	1 058	296	21
江 苏 Jiangsu	1 196	333	17
浙 江 Zhejiang	525	104	16
福 建 Fujian	304	130	7
山 东 Shandong	2 275	932	29
广 东 Guangdong	2 152	1 046	23
广 西 Guangxi	190	18	0
海 南 Hainan	78	3	2
其 他 Others	1 247	640	8

6-15 分地区海洋科研机构科技专利情况
Marine Scientific and Technological Patents of the Research Institutions by Region

单位：件 (unit)

地 区 Region	专利申请受理数 Number of Patent Applications Accepted	#发明专利 Patents for Discoveries	专利授权数 Number of Patents Granted	#发明专利 Patents for Discoveries	拥有发明专利总数 Total Number of Patents for Discoveries
合 计 Total	**6 111**	**4 813**	**4 020**	**2 647**	**13 966**
北 京 Beijing	2 371	2 095	1 582	1 236	7 027
天 津 Tianjin	134	72	96	28	191
河 北 Hebei	6	6	4	4	13
辽 宁 Liaoning	595	544	474	408	1 954
上 海 Shanghai	1 196	895	730	490	2 224
江 苏 Jiangsu	220	143	142	48	271
浙 江 Zhejiang	350	150	193	22	124
福 建 Fujian	60	49	20	12	273
山 东 Shandong	498	370	437	230	777
广 东 Guangdong	432	356	216	124	695
广 西 Guangxi	54	30	33	13	51
海 南 Hainan	0	0	6	3	13
其 他 Others	195	103	87	29	353

6-16　分地区海洋科研机构R & D情况
R & D in the Marine Scientific Research Institutions by Region

行　业 Industry	R & D人员 （人） R & D Personnel (person)	R & D经费内部支出 （千元） R & D Internal Expenditure (1 000 yuan)	R & D课题数 （项） Number of R & D Projects (item)
合 计 **Total**	**28 243**	**15 663 775**	**13 049**
北 京 Beijing	9 834	6 099 555	4 630
天 津 Tianjin	1 682	796 171	414
河 北 Hebei	268	68 450	63
辽 宁 Liaoning	822	570 547	259
上 海 Shanghai	2 659	2 163 707	729
江 苏 Jiangsu	2 070	1060 412	1 573
浙 江 Zhejiang	723	391 208	280
福 建 Fujian	762	354 156	518
山 东 Shandong	3 211	1 950 815	1 316
广 东 Guangdong	3 597	1 408 427	1 865
广 西 Guangxi	382	95 296	93
海 南 Hainan	0	0	0
其 他 Others	2 233	705 031	1 309

主要统计指标解释

1. 海洋科研机构 指有明确的研究方向和任务，有一定水平的学术带头人和一定数量、质量的研究人员，有开展研究工作的基本条件，长期有组织地从事海洋研究与开发活动的机构。

2. 从业人员 指由本机构年末直接组织安排工作并支付工资的各类人员总数。包括固定职工、国家有编制的合同制职工、招聘人员和返聘的离退休人员。不包括离退休人员、停薪留职人员。

3. 从事科技活动人员 指从业人员中的科技管理人员、课题活动人员和科技服务人员。

4. 高级职称 指研究员、副研究员；教授、副教授；高级工程师；高级农艺师；正、副主任医（药、护、技)师；高级实验师；高级统计师；高级经济师；高级会计师；编审(正、副编审)；译审(正、副译审)、高级(主任)记者；正、副研究馆员等。

5. 中级职称 指助理研究员；讲师；工程师；农艺师；主治医(药、护、技)师；实验师；统计师；经济师；会计师；编辑；翻译；记者；馆员等。

6. 初级职称 指研究实习员；助教；助理工程师、技术员；助理农艺师、农业技术员；医(药、护、技)师、医(药、护、技)士；助理实验师、实验员；助理统计师、统计员；助理经济师；助理会计师、会计员；助理编辑、见习编辑；助理翻译；助理记者；助理馆员、管理员等。

7. 科技经费筹集额 指从各种渠道筹集到的计划用于本单位科技活动的经费，不论来源渠道如何。

8. 政府资金 指由各级政府部门直接拨款或企事业单位利用政府资金委托本机构从事科学技术活动所获得的收入。

9. 生产经营活动收入 指本机构在科研、技术等专业业务活动以外开展非独立核算的经营活动取得的收入，包括产品（商品）销售收入、经营服务收入、工程承包收入、租赁收入和其他经营收入。

10. 其他收入 指开展科技活动与生产经营活动以外的各项活动的收入，包括：用于离退休人员的政府拨款。

11. 非科技活动借贷款 指本机构为开展非科技活动从各种渠道获得的各类借、贷款。不论偿还形式、期限和数额如何，均按当年获得的借、贷款额填报。不包括基本建设贷款。

12. 基础研究 为获得新知识而进行的独创性研究。其目的是揭示观察到的现象和事实的基本原理和规律，而不以任何特定的实际应用为目的。

13. 应用研究 为获得新的科学技术知识而进行的独创性研究。它主要针对某一特定的实际应用目的。应用研究通常是为了确定基础研究成果或知识的可能的用途，或是为达到某一具体的、预定的实际目的确定新的方法(原理性)或途径。

14. 试验发展 利用从研究或实际经验获得的知识，为生产新的材料、产品和装置，建立新的工艺和系统，以及对已生产或建立的上述各项进行实质性的改进而进行的系统性工作。

15. 成果应用 为解决 R&D 活动阶段产生的新产品、新装置、新工艺、新技术、新方法、新系统和服务等能投入生产或在实际应用中所存在的技术问题而进行的系统性活动。它不具有创新成分。此类活动包括为达到生产目的而进行的定型设计和试制以及为扩大新产品的生产规模和探索新方法、新技术、新工艺等的应用领域而进行的适应性试验。

16. 科技服务　与科学研究与实验发展有关，并有助于科学技术知识的产生、传播和应用的活动。包括为扩大科技成果的使用范围而进行的示范性推广工作；为用户提供科技信息和文献服务的系统性工作；为用户提供可行性报告、技术方案、建议及进行技术论证等技术咨询工作；自然、生物现象的日常观测、监测，资源的考查和勘探；有关社会、人文、经济现象的通用资料的收集，如统计、市场调查等以及这些资料的常规分析与整理；为社会和公众提供的测试、标准化、计量、计算、质量控制和专利服务，不包括工商企业为进行正常生产而开展的上述活动。

17. 生产性活动　由于业务特殊的工艺设备条件，或掌握某种技术专长或诀窍，所进行的小量非常规生产。

18. 科技论文　在全国性学报或学术刊物上、省部属大专院校对外正式发行的学报或学术刊物上发表的论文以及向国外发表的论文。

19. 科技著作　经过正式出版部门编印出版的科技专著、大专院校教科书、科普著作。

20. 专利申请受理数　当年本单位向专利管理部门提出申请并被受理的职务专利申请件数。

21. 专利授权数　当年由专利管理部门授予本单位专利权的职务专利件数。

Explanatory Notes on Main Statistical Indicators

1. Marine Scientific Research Institution refers to the institution which has definite research orientations and tasks, high-level academic leading personnel and fair-sized, qualified research personnel, and basic conditions for research work and which is engaged for a long time in the marine research and development activities in an organized way.

2. Employees refer to the total number of personnel of various kinds employed and paid by the institution at the end of the year, including fixed employees, contract workers of staff belonging to the state authorized staff , recruited personnel, reemployed retired personnel, but not including the retired and the personnel on leave with pay suspension.

3. Personnel Engaged in Scientific and Technological Activities refers to the personnel for scientific and Technological management, personnel engaged in the activities of research topics and scientific and technological service personnel.

4. Senior Technical Title refer to research scientist, associate research scientist; professor, associate professor; senior engineer; senior agronomist; professor-rank and associate professor-rank doctor (pharmacists, nurses and technicians); senior laboratory technician; senior statisticians; senior economic engineer; chief accountant; senior editor (professor and associate professor ranks); senior translator (professor and associate professor ranks); senior journalist; research librarian (professor and associate professor ranks), etc.

5. Intermediate Technical Title refer to assistant research scientist; lecturer; engineer; agronomist; lecturer-rank doctor (pharmacist, nurse and technician); laboratory technician; lecturer-rank statistician;

economic engineer; accountant; editor; translator; journalist; librarian, etc.

6. Primary Technical Title refers to trainee researcher; assistant; assistant engineer, technician, assistant agronomist, agricultural technician; assistant-rank doctor (pharmacist, nurse and technician); assistant laboratory technician; assistant statistician; assistant economic engineer, assistant accountant; assistant editor, editor on probation; assistant translator; assistant journalist; assistant research librarian, librarian ,etc.

7. Amount of Scientific and Technological Funds Raised refers to the funds raised through all channels planned to be used as funds for the scientific and technological activities in the institution regardless of their source and channels.

8. Funds from government refers to the direct appropriations by the government departments at all levels or the earnings from conducting scientific and technological activities entrusted to the institution by enterprises as institutions by earning the funds from government.

9. Earnings from Production as Business Activities refer to the incomes obtained from the non-independent accounting business activities carried out by the institution beyond the scientific research, technological and professional activities, including these from sale of products(goods), business and service, contracted projects, leasing and other business.

10. Other Incomes refer to those from the various activities conducted other than scientific and technological activities and production and business activities, including the government appropriations for retired personnel.

11. Loan for Non-scientific as Technological Activities refers to the various types loan obtained by the institutions through all channels for carrying out non-scientific and technological activities, not including the load for capital construction. The loan, irrespective of its form of reimbursement, term and amount is filled in a form and submitted to the authorities as the amount acquired in the current year.

12. Basic Research refers to the original research to acquire new knowledge. It is aimed at revealing the basic principles and laws of the phenomena and facts observed, but not at any specific practical applications.

13. Applied Research refers to the original research to acquire new scientific and technological knowledge. It mainly serves the purpose of a particular practical application. The purpose of applied research is usually to define the potential uses of the research finds or knowledge obtained from basic research or to identify new methods (principles) or ways to reach a specific and predetermined goal.

14. Experimental Development refers to the systematic work carried out to establish new technologies and systems for producing new materials, products and equipment by using the knowledge obtained from research or practical experience, or to make substantial improvement of the above-mentioned which have been produced or established.

15. Result Application refers to the systematic activities carried out to solve the technical problems that might crop up in the production or practical application of the new products, devices, technologies, techniques, methods, systems and service occurring in the course of R & D activities.

They do not bring forth new ideas. Such activities include the finalizing design and trial-production for the purpose of production as well as the adaptive tests to expand the production scale of new products and the application areas of new methods, techniques and technologies.

16. Scientific and Technological Service refers to the activities that are associated with the scientific research and experimental development, and contribute to the generation, dissemination and application of scientific and technological knowledge, which include the demonstrative work of popularization to enlarge the use scope of scientific and technological achievements; the systematic work of providing the users with scientific and technological information and literature service; the technical consultation work of providing users with feasibility reports, technical schemes and recommendations and carrying out technical demonstration; routine observation and monitoring of natural and biological phenomena, and the survey and exploration of resources; collection of universal data on the appropriate social, cultural and economic phenomena, such as statistics and market survey, as well as the routine analysis and sorting-out of these data; the provision for the society and the public of such service as testing, standardization, computation, quality control and patent, but not including the type of the above-mentioned activities carried out by industrial and commercial enterprises for the purpose of normal production.

17. Productive Activity refers to the small-scale and non-conventional productions due to the presence of special technologies and equipment or mastery of a particular technical expertise or secret of success.

18. Scientific Treatises refer to the theses published in the national journals or academic publications, those officially issued journals or academic publications by universities and colleges under provinces or ministries as well as theses published abroad.

19. Scientific and Technological Works refer to the scientific and technological monographs, textbooks for universities and colleges and popular science books published by the official publishing houses.

20. Number of Patents Applied and Accepted refers to the number of professional patent applications of the unit to the patent administrative department and accepted by it in the year.

21. Number of Patents Granted refers to the number of the professional patents granted to the unit by the patent administrative department in the year.

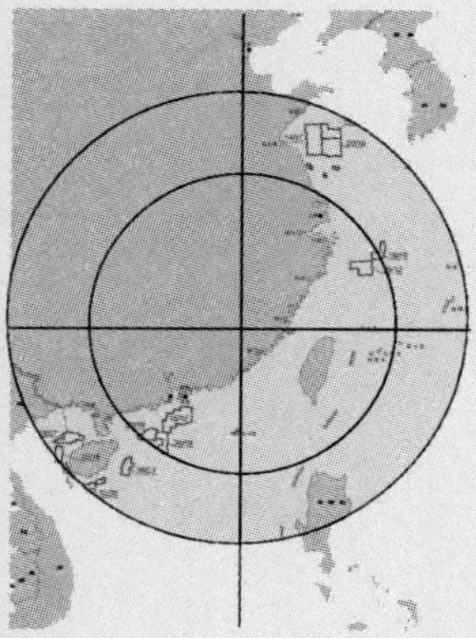

7 海洋教育
Marine Education

7-1 全国各海洋专业博士研究生情况
Doctoral Students from Marine Specialities

专业 Speciality	专业点数（个） Number of Speciality Agencies (unit)	学生数（人） Number of Students (person)			
		毕业生 Graduates	招生 Entrants	在校生 Enrollment	预计毕业生数 Estimated Graduates of Next Year
合计 Total	**140**	**672**	**939**	**4 251**	**2 160**
物理海洋学 Physical Oceanography	5	46	61	272	143
海洋化学 Marine Chemistry	4	24	36	157	82
海洋生物学 Marine Biology	8	59	91	298	111
海洋地质 Marine Geology	7	57	49	222	130
海洋科学学科 Marine Science Subjects	6	68	113	451	215
水生生物学 Hydrobiology	19	82	78	342	178
水文学及水资源 Hydrology and Water Resource	18	92	111	652	388
港口海岸及近海工程 Coastal Harbour and Offshore Engineering	10	31	55	286	164

7-1 续表 continued

专 业 Speciality	专业点数（个） Number of Speciality Agencies (unit)	学生数（人） Number of Students (person)			
		毕业生 Graduates	招 生 Entrants	在校生 Enrollment	预计毕业生数 Estimated Graduates of Next Year
船舶与海洋结构物设计制造 Ships and Marine Structures Design and Manufacture	12	67	57	380	224
轮机工程 Turbine Engineering	8	30	59	294	106
水声工程 Hydroacoustic Engineering	8	19	36	187	89
船舶与海洋工程学科 Ship and Ocean Engineering Subjects	6	13	57	215	99
水产品加工及贮藏工程 Aquatic Products Processing and Storing Engineering	9	11	6	43	31
水产养殖 Aquaculture	7	52	61	239	114
捕捞学 Science of Fishing	2	0	4	17	10
渔业资源 Fishery Resource	4	10	24	86	38
水产学科 Fishery Subjects	6	11	41	109	38
航空、航天与航海医学 Aeronautical, Aerospace and Nautical Medicine	1	0	0	1	0

7-2 全国各海洋专业硕士研究生情况
Postgraduate Students from Marine Specialities

专 业 Speciality	专业点数（个） Number of Speciality Agencies (unit)	学生数（人） Number of Students (person)			
		毕业生 Graduates	招 生 Entrants	在校生 Enrollment	预计毕业生数 Estimated Graduates of Next Year
合 计 Total	**332**	**3 031**	**3 478**	**10 318**	**3 403**
物理海洋学 Physical Oceanography	13	71	134	381	99
海洋化学 Marine Chemistry	16	104	143	437	144
海洋生物学 Marine Biology	23	331	292	850	282
海洋地质 Marine Geology	15	112	127	385	127
海洋科学学科 Marine Science Subjects	13	104	348	828	199
水生生物学 Hydrobiology	45	311	220	712	257
水文学及水资源 Hydrology and Water Resource	52	378	440	1 231	382
港口海岸及近海工程 Coastal Harbour and Offshore Engineering	22	284	253	800	293

7-2 续表 continued

专 业 Speciality	专业点数（个） Number of Speciality Agencies (unit)	学生数（人） Number of Students (person)			
		毕业生 Graduates	招 生 Entrants	在校生 Enrollment	预计毕业生数 Estimated Graduates of Next Year
船舶与海洋结构物设计制造 Ships and Marine Structures Design and Manufacture	20	333	279	968	405
轮机工程 Turbine Engineering	13	183	236	707	252
水声工程 Hydroacoustic Engineering	9	94	128	365	126
船舶与海洋工程学科 Ship and Ocean Engineering Subjects	16	101	245	648	154
水产品加工及贮藏工程 Aquatic Products Processing and Storing Engineering	24	94	67	266	104
水产养殖 Aquaculture	28	391	402	1 231	432
捕捞学 Science of Fishing	5	31	19	72	27
渔业资源 Fishery Resource	10	82	71	236	83
水产学科 Fishery Subjects	6	11	74	187	23
航空、航天与航海医学 Aeronautical, Aerospace and Nautical Medicine	2	16		14	14

7-3 全国普通高等教育各海洋专业本科学生情况
Undergraduates from Marine Specialities in the National Ordinary Higher Education

专 业 Speciality	专业点数（个） Number of Speciality Agencies (unit)	学生数（人） Number of Students (person)			
		毕业生 Graduates	招 生 Entrants	在校生 Enrollment	预计毕业生数 Estimated Graduates of Next Year
合 计 Total	**251**	**14 466**	**18 250**	**67 959**	**16 284**
海洋科学 Marine Science	25	945	987	3 840	948
海洋技术(注：可授理学或工学学士学位) Marine Technology (Note: It may confer bachelor's degrees in science and engineering)	20	599	621	2 723	736
海洋资源与环境 Marine Living Resources and Environment	11	110	410	1 155	144
海洋科学类专业 New Specialties Under the Category of Marine Sciences	7	23	744	1 268	96
港口航道与海岸工程 Harbour Channel and Coastal Engineering	29	1 860	1 653	7 397	2 066
航海技术 Nautical Technology	17	2 000	2 741	10 247	2 527
轮机工程 Turbine Engineering	22	2 600	3 218	12 553	3 109
船舶与海洋工程 Ship and Marine Engineering	30	3 032	2 316	10 945	3 078
海洋工程与技术 Ocean Engineering and Technology	3	20	57	305	76
海洋资源开发技术 Marine Resources Exploitation Technology	8	0	259	809	90
海洋工程类专业 New Speciality of Marine Engineering	3	0	486	684	0
水产养殖学 Aquaculture	55	2 597	3 520	12 276	2 642
海洋渔业科学与技术 Science and Technology of Marine Fishery	9	364	579	1 926	394
水族科学与技术 Science and Technology of Aquatic Animals	9	194	420	1 243	255
水产类专业 Fishery-type Specialities	1	0	128	128	0
海事管理 Maritime Affairs Manegement	2	122	111	460	123

7-4 全国普通高等教育各海洋专业专科学生情况
Students from Marine Specialities of the Colleges for Professional Training in the National Ordinary Higher Education

专 业 Speciality	专业点数（个） Number of Speciality Agencies (unit)	学生数（人） Number of Students (person)			
		毕业生 Graduates	招 生 Entrants	在校生 Enrollment	预计毕业生数 Estimated Graduates of Next Year
合 计 Total	**516**	**33 745**	**27 058**	**87 517**	**30 881**
水产养殖技术 Aquaculture Technology	36	1 165	1 130	3 406	1 259
水生动植物保护 Aquatic Animals and Plants Conservation	2	17	0	0	0
渔业综合技术 Integrated Fishery Technology	1	0	0	8	0
水产养殖类专业 Aquaculture Type of Majors	1	0	35	35	0
水运管理 Water Transport Management	3	138	45	249	114
国际航运业务管理 International Shipping Business Management	35	1 929	2 201	6 644	2 181
轮机工程技术 Engine Engineering Technology	50	6 004	4 124	13 583	4 942
船舶工程技术 Ship Engineering Technology	42	5 647	3 819	12 518	4 721
船舶检验 Ship Inspection	7	174	166	341	68
船机制造与维修 Ship Engines Manufacturing and Maintenance	7	468	387	1 151	391
船舶舾装 Ship Equipment and Installations	4	180	92	296	104
水上运输类专业 Water Transportation Type of Majors	1	0	0	1	1
港口业务管理 Harbour Business Management	20	1 155	1 114	2 868	902

7-4 续表 continued

专 业 Speciality	专业点数（个） Number of Speciality Agencies (unit)	学生数（人） Number of Students (person)			
		毕业生 Graduates	招 生 Entrants	在校生 Enrollment	预计毕业生数 Estimated Graduates of Next Year
港口物流设备与自动控制 Harbour Logistics Equipment and Automatic Control	20	1 009	864	2 831	1 044
集装箱运输管理 Container Transportation Management	18	967	613	2 046	713
港口工程技术 Harbour Engineering	12	523	385	1 460	592
报关与国际货运 Customs Clearing and International Freight Transport	216	13 248	10 348	35 588	12 629
港口与航运管理 Harbour and Shipping Management	11	434	650	1 698	462
港口机械应用技术 Applied Harbour Machinery Technology	4	213	257	733	283
港口运输类专业 Port Transportation Type of Majors	3	0	73	88	0
水文自动化测报技术 New Specialties Under the Category of Marine Sciences	1	0	36	82	18
水信息技术 Hydrological Information Technology	2	40	5	13	8
水政水资源管理 Water Resources Management by Water Administration	4	95	163	489	105
港口航道与治河工程 Harbour, Channel and River Harnessing Engineering	4	138	135	354	101
水环境监测与保护 Water Environmental Monitoring and Protection	6	187	283	726	146
海关管理 Customs Management	6	14	133	309	97

7-5 全国成人高等教育各海洋专业本科学生情况
Undergraduates from Marine Specialities in the National Adult Higher Education

专 业 Speciality	专业点数（个） Number of Speciality Agencies (unit)	学生数（人） Number of Students (person)			
		毕业生 Graduates	招 生 Entrants	在校生 Enrollment	预计毕业生数 Estimated Graduates of Next Year
合 计 **Total**	**57**	**2 126**	**1 526**	**5 028**	**2 268**
海洋技术 （注：可授理学或工学学士学位） Marine Technology (Note: It may confer bachelor's degrees in science and engineering)	1	4	34	127	18
港口航道与海岸工程 Harbour Channel and Coastal Engineering	3	58	17	60	26
航海技术 Navigation Technology	9	156	126	594	326
轮机工程 Turbine Engineering	10	292	230	819	367
船舶与海洋工程 Ship and Marine Engineering	14	1 480	909	2 793	1 317
水产养殖学 Aquaculture	19	135	210	635	214
海事管理 Maritime Affairs Management	1	1	0	0	0

7-6 全国成人高等教育各海洋专业专科学生情况
Students from Marine Specialities of the Colleges for Professional Training in the National Adult Higher Education

专 业 Speciality	专业点数（个） Number of Speciality Agencies (unit)	学生数（人） Number of Students (person)			
		毕业生 Graduates	招 生 Entrants	在校生 Enrollment	预计毕业生数 Estimated Graduates of Next Year
合 计 Total	**144**	**6 524**	**3 776**	**13 087**	**7 030**
水产养殖技术 Aquaculture Technology	23	167	843	1 181	143
渔业综合技术 Fishery Ingrated Technology	1	1	2	141	0
水产养殖类新专业 New Specialties under the Category of Aquaculture	2	6	9	13	4
国际航运业务管理 International Shipping Business Management	3	72	73	297	182
轮机工程技术 Engine Engineering	33	3 430	1 066	5 829	3 714
船舶工程技术 Ship Engineering	24	1 702	1 152	3 485	1 822
船舶检验 Ship Inspection	2	7	19	30	7
船机制造与维修 Ship Engines Manufacturing and Maintenance	2	61	6	121	111
水上运输类新专业 New Specialties under the Category of Water Transport	2	35	0	73	73

7-6 续表 continued

专 业 Speciality	专业点数（个） Number of Speciality Agencies (unit)	学生数（人） Number of Students (person)			
		毕业生 Graduates	招 生 Entrants	在校生 Enrollment	预计毕业生数 Estimated Graduates of Next Year
港口业务管理 Harbour Business Management	1	0	1	1	0
港口物流设备与自动控制 Harbour Logistics Equipment and Automatic Control	1	0	50	148	50
港口工程技术 Harbour Engineering	1	0	0	1	1
报关与国际货运 Customs Declaration/International Freight Transport	37	970	371	1 513	865
港口与航运管理 Harbour and Shipping Management	3	0	149	181	32
水文自动化测报技术 Automated Hydrographic Measuring and Reporting Technology	1	0	1	1	0
水政水资源管理 Water Resources Management by Water Administration	4	24	16	54	26
港口航道与治河工程 Harbour Channel and River Harnessing Projects	1	0	10	10	0
海关管理 Customs Management	2	1	8	8	0
船艇动力管理 Ship and Boat Power Equipment Management	1	48	0	0	0

7-7 全国中等职业教育各海洋专业学生情况
Students from Marine Specialities in the National Secondary Vocational Education

专业 Speciality	专业点数（个） Number of Speciality Agencies (unit)	学生数（人） Number of Students (person)			
		毕业生 Graduates	招生 Entrants	在校生 Enrollment	预计毕业生数 Estimated Graduates of Next Year
合　计　Total	**398**	**24 193**	**16 728**	**44 899**	**17 243**
海水生态养殖 Seawater Ecological Cultivation	20	1 827	764	2 252	1 089
航海捕捞 Sea Fishing	4	160	0	41	41
农林牧渔类新专业 New Specialities of Agriculture, Forestry, Animal Husbandry and Fishery	48	4 389	3 555	8 319	3 033
水文与水资源勘测 Hydrological and Water Resources Survey	6	186	130	267	86
风电场机电设备运行与维护 Operation and Maintenance of Electromechanical Equipment in	42	2 089	1 473	3 719	1 346
船舶制造与修理 Ships Building and Repair	55	3 322	2 813	7 744	2 140
船舶机械装置安装与维修 Installation and Maintenance of Ships' Mechanical Equipment	13	561	523	1 396	577
船舶驾驶 Ship Piloting	70	5 358	3 299	9 123	4 358

7-7 续表 continued

专 业 Speciality	专业点数（个） Number of Speciality Agencies (unit)	学生数（人） Number of Students (person)			
		毕业生 Graduates	招 生 Entrants	在校生 Enrollment	预计毕业生数 Estimated Graduates of Next Year
轮机管理 Engines Management	57	3 629	2 174	5 493	2 523
船舶水手与机工 Ship Sailors and Mechanics	21	760	313	1 643	322
船舶电气技术 Ship Electric Technology	24	602	597	1 269	335
船舶通信与导航 Marine Communication and Navigation	1	0	0	2	0
外轮理货 Foreign Ships Freight Forwarding	11	177	190	788	317
船舶检验 Ships Inspection	3	313	30	89	47
港口机械运行与维护 Operation and Maintenance of Harbour Machinery	21	798	834	2 671	1 022
工程潜水 Engineering Diving	2	22	33	83	7

注：此表不包含技工学校相关数据。

Note: Data related to Technical Schools are not included in the table.

7-8 分地区各海洋专业博士研究生情况
Doctoral Students in Marine Specialities by Regions

地区 Region	专业点数（个） Number of Speciality Agencies (unit)	学生数（人） Number of Students (person)			
		毕业生 Graduates	招生 Entrants	在校生 Enrollment	预计毕业生数 Estimated Graduates of Next Year
合计 Total	**140**	**672**	**939**	**4 251**	**2 160**
北京 Beijing	13	164	199	644	233
天津 Tianjin	4	7	5	23	13
辽宁 Liaoning	8	39	55	340	155
上海 Shanghai	18	56	86	417	236
江苏 Jiangsu	13	51	82	491	263
浙江 Zhejiang	5	1	26	89	19
福建 Fujian	8	17	56	178	81
山东 Shandong	13	165	182	933	561
广东 Guangdong	10	23	28	130	65
广西 Guangxi	1	0	1	2	0
海南 Hainan	1	0	2	5	1
其他 Others	46	149	217	999	533

7-9 分地区各海洋专业硕士研究生情况
Postgraduate Students in Marine Specialities by Regions

地 区 Region	专业点数（个） Number of Speciality Agencies (unit)	学生数（人） Number of Students (person)			
		毕业生 Graduates	招 生 Entrants	在校生 Enrollment	预计毕业生数 Estimated Graduates of Next Year
合 计 Total	**332**	**3 031**	**3 478**	**10 318**	**3 403**
北 京 Beijing	25	172	281	768	226
天 津 Tianjin	8	57	83	235	66
河 北 Hebei	7	17	41	95	22
辽 宁 Liaoning	24	272	335	1 086	398
上 海 Shanghai	30	378	392	1 158	400
江 苏 Jiangsu	25	348	378	1 120	374
浙 江 Zhejiang	26	166	274	683	189
福 建 Fujian	22	183	173	574	191
山 东 Shandong	32	356	384	1 175	403
广 东 Guangdong	18	172	163	563	204
广 西 Guangxi	4	19	13	45	18
海 南 Hainan	3	29	21	67	22
其 他 Others	108	862	940	2 749	890

7-10 分地区普通高等教育各海洋专业本科学生情况
Undergraduates in the Marine Specialities of Ordinary Higher Education by Regions

地 区 Region	专业点数（个） Number of Speciality Agencies (unit)	学生数（人） Number of Students (person)			
		毕业生 Graduates	招 生 Entrants	在校生 Enrollment	预计毕业生数 Estimated Graduates of Next Year
合 计 Total	**251**	**14 466**	**18 250**	**67 959**	**16 284**
北 京 Beijing	3	90	92	423	156
天 津 Tianjin	12	847	798	3 404	864
河 北 Hebei	10	211	482	1 332	176
辽 宁 Liaoning	23	2 157	2 352	8 900	2 294
上 海 Shanghai	17	884	1 390	5 528	1 566
江 苏 Jiangsu	29	1 383	1 456	5 882	1 534
浙 江 Zhejiang	30	745	1 080	4 086	900
福 建 Fujian	14	1 034	1 521	5 865	1 266
山 东 Shandong	30	1 658	2 157	8 239	1 866
广 东 Guangdong	17	860	2 117	5 583	901
广 西 Guangxi	6	178	248	941	190
海 南 Hainan	2	120	140	556	120
其 他 Others	58	4 299	4 417	17 220	4 451

7-11 分地区普通高等教育各海洋专业专科学生情况
Students from of the Marine Specialities Colleges for Professional Training in the Ordinary Higher Education by Regions

地　区 Region	专业点数（个） Number of Speciality Agencies (unit)	学生数（人） Number of Students (person)			
		毕业生 Graduates	招生 Entrants	在校生 Enrollment	预计毕业生数 Estimated Graduates of Next Year
合　计 **Total**	**516**	**33 745**	**27 058**	**87 517**	**30 881**
北　京 Beijing	1	0	39	60	0
天　津 Tianjin	14	1 440	1 202	3 692	1 317
河　北 Hebei	31	856	548	1 749	589
辽　宁 Liaoning	37	2 966	2 800	9 769	3 714
上　海 Shanghai	29	2 029	1 362	4 868	1 840
江　苏 Jiangsu	58	4 826	3 015	11 200	4 182
浙　江 Zhejiang	34	1 999	1 572	5 712	2 034
福　建 Fujian	33	1 689	1 437	4 722	1 632
山　东 Shandong	69	5 779	4 649	14 285	4 950
广　东 Guangdong	34	2 026	1 939	6 148	1 933
广　西 Guangxi	27	1 204	1 245	3 526	1 169
海　南 Hainan	7	328	440	1 206	437
其　他 Others	142	8 603	6 810	20 580	7 084

7-12　分地区成人高等教育各海洋专业本科学生情况
Students from Marine Specialities in the Adult Higher Education by Regions

地　区 Region	专业点数（个） Number of Speciality Agencies (unit)	学生数（人） Number of Students (person)			
		毕业生 Graduates	招　生 Entrants	在校生 Enrollment	预计毕业生数 Estimated Graduates of Next Year
合　计 Total	**57**	**2 126**	**1 526**	**5 028**	**2 268**
天　津 Tianjin	2	75	32	128	47
河　北 Hebei	1	18	4	24	9
辽　宁 Liaoning	7	202	137	594	457
上　海 Shanghai	4	51	103	439	122
江　苏 Jiangsu	4	979	712	2 069	926
浙　江 Zhejiang	6	48	56	312	103
福　建 Fujian	1	7	0	34	16
山　东 Shandong	10	97	56	178	74
广　东 Guangdong	5	253	88	435	185
广　西 Guangxi	1	0	1	1	0
其　他 Others	16	396	337	814	329

7-13 分地区成人高等教育各海洋专业专科学生情况
Students from Marine Specialities of the colleges for Professional Training in the Adult Higher Education by Regions

地 区 Region	专业点数（个） Number of Speciality Agencies (unit)	学生数（人） Number of Students (person)			
		毕业生 Graduates	招 生 Entrants	在校生 Enrollment	预计毕业生数 Estimated Graduates of Next Year
合 计 Total	**144**	**6 524**	**3 776**	**13 087**	**7 030**
天 津 Tianjin	3	132	28	266	184
河 北 Hebei	1	10	1	4	0
辽 宁 Liaoning	17	1 386	817	3 206	2 133
上 海 Shanghai	11	738	239	1 254	489
江 苏 Jiangsu	20	1 332	967	2 898	1 588
浙 江 Zhejiang	10	1 029	109	568	456
福 建 Fujian	4	221	598	777	159
山 东 Shandong	16	410	108	608	392
广 东 Guangdong	15	58	136	772	439
广 西 Guangxi	4	4	12	16	4
其 他 Others	43	1 204	761	2 718	1 186

7-14 分地区中等职业教育各海洋专业学生情况
Students from Marine Specialities in the Secondary Vocational Education by Regions

地区 Region	专业点数（个）Number of Speciality Agencies (unit)	学生数（人）Number of Students (person)			
		毕业生 Graduates	招生 Entrants	在校生 Enrollment	预计毕业生数 Estimated Graduates of Next Year
合计 Total	**398**	**24 193**	**16 728**	**44 899**	**17 243**
天津 Tianjin	7	994	608	1 436	557
河北 Hebei	23	1 781	1 044	3 399	1 501
辽宁 Liaoning	43	2 499	561	3 318	1 990
上海 Shanghai	16	1 338	1 271	3 465	1 359
江苏 Jiangsu	43	2 611	726	2 581	890
浙江 Zhejiang	21	1 353	1 420	2 929	946
福建 Fujian	35	1 917	3 012	5 074	1 690
山东 Shandong	58	3 672	2 060	6 342	3 014
广东 Guangdong	17	1 115	395	1 131	295
广西 Guangxi	9	366	651	2 324	884
海南 Hainan	4	45	96	230	73
其他 Others	122	6 502	4 884	12 670	4 044

7-15 分地区开设海洋专业高等学校教职工数
Number of Teaching and Administrative Staff in the Universities and Colleges Offering Marine Specialities by Regions

地　区 Region	学校（机构）数（个） Number of Colleges (Institutions) (unit)	教职工数（人） Number of Teaching and Administrative Staff (person)	专任教师数（人） Number of Full-Time Teachers (person)
合　计 Total	**408**	**478 025**	**306 669**
北　京 Beijing	4	5 725	3 536
天　津 Tianjin	12	15 957	11 020
河　北 Hebei	24	22 297	14 112
辽　宁 Liaoning	24	20 631	13 131
上　海 Shanghai	15	20 236	9 667
江　苏 Jiangsu	45	61 557	39 329
浙　江 Zhejiang	23	29 464	18 173
福　建 Fujian	22	22 988	14 423
山　东 Shandong	47	53 706	36 452
广　东 Guangdong	24	35 364	22 562
广　西 Guangxi	16	14 743	10 248
海　南 Hainan	6	6 892	3 967
其　他 Others	146	168 465	110 049

主要统计指标解释

海洋专业 指高等教育和中等职业教育所设的与海洋有关的专业。

Explanatory Notes on Main Statistical Indicators

Marine Speciality refers to the ocean-related speciality in higher education and the professional secondary vocational education.

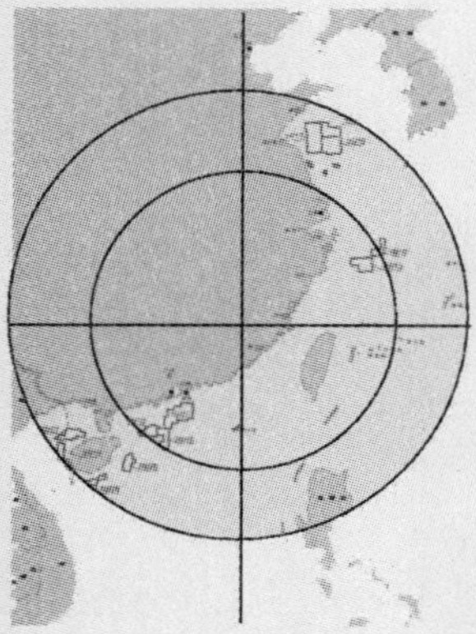

8 海洋环境保护与防灾减灾

Marine Environmental Protection and Disaster Mitigation

8-1 海区海水水质评价结果
Seawater Quality Assessment Results

项　目 Item	季　节 Season	合计 （平方千米） Total (km^2)	第二类水质海域面积 （平方千米） Area of the Second Grade Sea Waters (km^2)	第三类水质海域面积 （平方千米） Area of the Third Grade Sea Waters (km^2)
全海域 Total Area	春季	**167 810**	60 000	35 790
	夏季	**148 710**	43 280	42 740
	秋季	**217 980**	100 400	36 710
渤　海 Bohai Sea	春季	**35 880**	17 710	7 470
	夏季	**24 300**	8 180	6 600
	秋季	**54 530**	38 720	6 190
黄　海 Yellow Sea	春季	**35 380**	16 390	10 230
	夏季	**34 010**	12 510	13 540
	秋季	**49 160**	27 310	10 450
东　海 East China Sea	春季	**73 610**	19 020	9 210
	夏季	**67 700**	17 470	10 700
	秋季	**93 420**	27 220	11 330
南　海 South China Sea	春季	**22 940**	6 880	8 880
	夏季	**22 700**	5 120	11 900
	秋季	**20 870**	7 150	8 740

8-1 续表 continued

项 目 Item	第四类水质海域面积（平方千米） Area of the Fourth Grade Sea Waters (km^2)	劣于第四类水质海域面积（平方千米） Sea Area of the Sea Waters Inferior to the Fourth Grade (km^2)	首要超标污染物 Prime Pollutants Exceeding the Set Standard
全海域 Total Area	19 740	52 280	无机氮、活性磷酸盐、石油类 Inorganic Nitrogen, Active Phosphate, Oils
	21 550	41 140	
	23 510	57 360	
渤 海 Bohai Sea	4 540	6 160	无机氮、石油类 Inorganic Nitrogen, Oils
	3 770	5 750	
	3 620	6 000	
黄 海 Yellow Sea	6 050	2 710	无机氮、活性磷酸盐、石油类 Inorganic Nitrogen, Active Phosphate, Oils
	4 990	2 970	
	7 140	4 260	
东 海 East China Sea	7 580	37 800	无机氮、活性磷酸盐 Inorganic Nitrogen, Active Phosphate
	11 200	28 330	
	11 670	43 200	
南 海 South China Sea	1 570	5 610	无机氮、活性磷酸盐、石油类 Inorganic Nitrogen, Active Phosphate, Oils
	1 590	4 090	
	1 080	3 900	

8-2 海区废弃物海洋倾倒情况
Ocean Dumping of Wastes by Sea Area

海 区 Sea Area	疏浚物 （万立方米） Dredged Materials (10 000m^3)	惰性无机地质废料 （立方米） Inert, Inorganic Geologic Wastes (m^3)
合 计 **Total**	**12 546.5**	**1 303**
渤黄海 Bohai and Yellow Sea		635
东 海 East China Sea	7 685.0	
南 海 South China Sea	4 861.5	668

8-3 海区海洋石油勘探开发污染物排放入海情况
Discharge of Pollutants into the Sea from Offshore Oil Exploration and Exploitation

海 区 Sea Area	生产污水 （万立方米） Production Sewage (10 000m^3)	泥浆 （立方米） Sludge (m^3)	钻屑 （立方米） Debris from Drilling (m^3)	机舱污水 （立方米） Sewage from Engineroom (m^3)	食品废弃物 （立方米） Food Wastes (m^3)	生活污水 （立方米） Domestic Sewage (m^3)
合 计 Total	**15 644.2**	**37 782.7**	**32 158.1**	**3 207.5**	**612.6**	**323 022.7**
渤 海 Bohai Sea	214.0	2 474.0	7 888.0		0.6	34 849.0
黄 海 Yellow Sea						
东 海 East China Sea	93.3	1 610.0	5 033.3			5.4
南 海 South China Sea	15 337.0	33 698.7	19 236.8	3 207.5	612.0	288 168.3

8-4 沿海地区工业废水排放及处理情况
Discharge and Treatment of Industrial Waste Water by Coastal Regions

单位：万吨 (10 000 t)

地 区 Region		工业废水排放总量 Total Volume of Industrial Waste Water Discharged	直排入海 Discharged Directly into the Sea	工业废水处理量 Industrial Waste Water Treated
总 计 Total		**1 156 933.7**	**109 154.6**	**2 964 638.7**
环渤海经济区 Round-the-Bohai Sea Economic Zone	**合 计 Total**	**398 226.3**	**39 013.3**	**1 523 726.5**
	辽 宁 Liaoning	90 630.8	28 579.4	317 134.8
	河 北 Hebei	108 562.0	1 084.2	788 496.3
	天 津 Tianjin	19 011.3	43.4	32 529.7
	山 东 Shandong	180 022.3	9 306.2	385 565.6
长江三角洲经济区 Yangtze River Delta Economic Zone	**合 计 Total**	**398 209.6**	**17 963.8**	**726 221.4**
	江 苏 Jiangsu	204 890.0	778.3	407 334.1
	上 海 Shanghai	43 939.2	10 180.2	63 561.4
	浙 江 Zhejiang	149 380.4	7 005.3	255 325.9
海峡西岸经济区 Economic Zone on the West Side of the Taiwan Straits	**合 计 Total**	**102 051.7**	**42 888.9**	**163 571.7**
	福 建 Fujian	102 051.7	42 888.9	163 571.7
珠江三角洲经济区 Zhujiang River Delta Economic Zone	**合 计 Total**	**177 554.0**	**2 784.2**	**334 111.6**
	广 东 Guangdong	177 554.0	2 784.2	334 111.6
环北部湾经济区 Round-the-Beibu Gulf Economic Zone	**合 计 Total**	**80 892.1**	**6 504.5**	**217 007.5**
	广 西 Guangxi	72 936.3	2 898.9	208 938.8
	海 南 Hainan	7 955.8	3 605.6	8 068.7

8-5 沿海城市工业废水排放及处理情况
Discharge and Treatment of Industrial Waste Water by Coastal Cities

单位：万吨 (10 000 t)

沿海城市 Coastal City	工业废水排放总量 Total Volume of Industrial Waste Water Discharged	直排入海 Discharged Directly into the Sea	工业废水处理量 Industrial Waste Water Treated
天　津　Tianjin	19 011.3	43.4	32 529.7
唐　山　Tangshan	13 973.5	1 080.5	396 556.9
秦皇岛　Qinhuangdao	6 273.1	0.0	16 826.7
沧　州　Cangzhou	9 490.4	3.8	10 491.1
大　连　Dalian	40 150.3	27 086.4	6 503.2
丹　东　Dandong	4 327.1	1.3	4 067.9
锦　州　Jinzhou	3 353.4	20.5	3 388.6
营　口　Yingkou	2 794.7	872.3	3 436.0
盘　锦　Panjin	4 418.3	0.0	3 616.2
葫芦岛　Huludao	2 530.2	599.1	3 835.1
上　海　Shanghai	43 939.2	10 180.2	63 561.4
南　通　Nantong	15 808.9	0.0	13 739.2
连云港　Lianyungang	6 204.5	311.2	18 802.4
盐　城　Yancheng	17 472.1	442.0	21 385.0

8-5 续表1 continued

沿海城市 Coastal City	工业废水排放总量 Total Volume of Industrial Waste Water Discharged	直排入海 Discharged Directly into the Sea	工业废水处理量 Industrial Waste Water Treated
杭　州 Hangzhou	35 370.2	0.0	120 074.4
宁　波 Ningbo	16 545.6	5 578.5	35 313.8
温　州 Wenzhou	6 020.4	2.4	5 876.0
嘉　兴 Jiaxing	20 635.9	169.5	25 245.6
绍　兴 Shaoxing	26 341.2	0.0	26 508.2
舟　山 Zhoushan	2 150.4	1 126.6	955.1
台　州 Taizhou	6 065.0	128.4	5 026.9
福　州 Fuzhou	4 680.8	385.8	16 935.8
厦　门 Xiamen	27 380.4	22 558.6	28 785.4
莆　田 Putian	2 633.2	1 001.5	2 491.3
泉　州 Quanzhou	19 257.9	2 308.1	25 434.3
漳　州 Zhangzhou	23 963.3	16 589.7	22 476.9
宁　德 Ningde	1 496.8	33.4	1 662.6
青　岛 Qingdao	10 989.4	2 475.2	47 501.2
东　营 Dongying	8 624.1	0.0	28 827.9
烟　台 Yantai	9 180.6	2 469.3	16 068.0
潍　坊 Weifang	27 101.3	0.0	41 959.7
威　海 Weihai	2 710.1	84.6	3 164.6
日　照 Rizhao	7 542.2	4 277.1	8 127.3
滨　州 Binzhou	15 622.3	0.0	13 428.8

8-5 续表2 continued

沿海城市 Coastal City	工业废水排放总量 Total Volume of Industrial Waste Water Discharged	直排入海 Discharged Directly into the Sea	工业废水处理量 Industrial Waste Water Treated
广　州 Guangzhou	22 443.8	0.0	22 170.9
深　圳 Shenzhen	16 313.5	609.6	16 341.2
珠　海 Zhuhai	4 936.0	25.6	4 437.0
汕　头 Shantou	5 516.8	20.6	5 966.7
江　门 Jiangmen	17 284.0	71.4	19 351.9
湛　江 Zhanjiang	6 557.6	104.7	6 386.3
茂　名 Maoming	4 298.9	927.5	5 859.8
惠　州 Huizhou	8 465.2	806.8	7 822.3
汕　尾 Shanwei	1 920.0	4.8	588.4
阳　江 Yangjiang	2 306.2	213.2	1 631.0
东　莞 Dongguan	28 395.6	0.0	38 725.4
中　山 Zhongshan	8 071.7	0.0	8 797.2
潮　州 Chaozhou	2 795.9	0.0	1 358.3
揭　阳 Jieyang	3 847.6	0.0	3 152.2
北　海 Beihai	1 920.4	446.9	25 303.3
防城港 Fangchenggang	1 731.6	218.5	4 528.6
钦　州 Qinzhou	3 982.5	2 230.4	4 207.7
海　口 Haikou	775.9	3.4	690.1
三　亚 Sanya	46.1	3.2	40.8

8-6 沿海地带工业废水排放及处理情况
Discharge and Treatment of Industrial Waste Water by Coastal Counties

单位：万吨 (10 000 t)

地区 Region	工业废水排放总量 Total Volume of Industrial Waste Water Discharged	直排入海 Discharged Directly into the Sea	工业废水处理量 Industrial Waste Water Treated
天　津 Tianjin	**8 737.4**	**43.4**	**11 002.0**
河　北 Hebei	**6 629.1**	**3.8**	**54 994.3**
唐　山 Tangshan	1 953.3	0.0	43 640.4
秦皇岛 Qinhuangdao	3 780.4	0.0	10 519.6
沧　州 Cangzhou	895.4	3.8	834.3
辽　宁 Liaoning	**26 228.1**	**13 762.8**	**13 806.9**
大　连 Dalian	19 078.1	12 290.2	4 874.1
丹　东 Dandong	1 159.0	1.3	549.3
锦　州 Jinzhou	878.2	0.0	1 492.3
营　口 Yingkou	1 975.6	872.3	2 779.8
盘　锦 Panjin	759.6	0.0	681.2
葫芦岛 Huludao	2 377.7	599.1	3 430.2
上　海 Shanghai	**30 062.5**	**10 180.2**	**54 326.8**
江　苏 Jiangsu	**23 329.1**	**753.2**	**39 728.8**
南　通 Nantong	8 067.9	0.0	6 710.7
连云港 Lianyungang	4 389.0	311.2	17 383.1
盐　城 Yancheng	10 872.2	442.0	15 635.0

8-6 续表1 continued

地 区 Region	工业废水排放总量 Total Volume of Industrial Waste Water Discharged	直排入海 Discharged Directly into the Sea	工业废水处理量 Industrial Waste Water Treated
浙 江 Zhejiang	**63 210.4**	**6 702.7**	**73 851.2**
杭 州 Hangzhou	11 483.0	0.0	14 528.6
宁 波 Ningbo	14 685.3	5 445.1	24 405.6
温 州 Wenzhou	4 710.4	2.4	4 340.8
嘉 兴 Jiaxing	7 438.1	0.2	8 749.6
绍 兴 Shaoxing	18 423.8	0.0	17 202.7
舟 山 Zhoushan	2 150.4	1 126.6	955.1
台 州 Taizhou	4 319.3	128.4	3 668.8
福 建 Fujian	**71 876.0**	**42 126.2**	**76 091.0**
福 州 Fuzhou	3 553.5	385.8	15 320.4
厦 门 Xiamen	27 380.4	22 558.6	28 785.4
莆 田 Putian	1 877.6	250.7	1 698.0
泉 州 Quanzhou	17 229.0	2 308.1	22 583.9
漳 州 Zhangzhou	20 795.0	16 589.7	6 701.1
宁 德 Ningde	1 040.6	33.4	1 002.3
山 东 Shandong	**43 194.2**	**5 369.6**	**120 062.8**
青 岛 Qingdao	9 489.2	2 475.2	45 899.8
东 营 Dongying	8 195.4	0.0	28 531.1
烟 台 Yantai	7 540.2	2 469.3	14 567.9
潍 坊 Weifang	11 492.9	0.0	24 404.0
威 海 Weihai	1 997.8	82.9	2 469.7
日 照 Rizhao	847.1	342.3	1 651.4
滨 州 Binzhou	3 631.6	0.0	2 539.0

8-6 续表2 continued

地 区 Region	工业废水排放总量 Total Volume of Industrial Waste Water Discharged	直排入海 Discharged Directly into the Sea	工业废水处理量 Industrial Waste Water Treated
广 东 Guangdong	**94 617.2**	**875.9**	**106 647.8**
广 州 Guangzhou	12 050.8	0.0	12 439.4
深 圳 Shenzhen	10 802.7	472.2	10 206.8
珠 海 Zhuhai	2 457.4	0.0	2 069.0
汕 头 Shantou	5 479.2	20.6	5 960.9
江 门 Jiangmen	12 340.6	71.4	14 345.3
湛 江 Zhanjiang	6 352.8	93.7	6 235.4
茂 名 Maoming	1 824.8	0.0	3 450.4
惠 州 Huizhou	1 790.4	0.0	1 679.5
汕 尾 Shanwei	1 781.2	4.8	562.1
阳 江 Yangjiang	1 471.8	213.2	922.5
东 莞 Dongguan	28 395.6	0.0	38 725.4
中 山 Zhongshan	8 071.7	0.0	8 797.2
潮 州 Chaozhou	1 277.9	0.0	606.2
揭 阳 Jieyang	520.2	0.0	647.6
广 西 Guangxi	**5 560.8**	**2 895.8**	**30 819.0**
北 海 Beihai	1 920.4	446.9	25 303.3
防城港 Fangchenggang	1 161.3	218.5	2 894.4
钦 州 Qinzhou	2 479.0	2 230.4	2 621.3
海 南 Hainan	**732.2**	**0.0**	**654.2**
海 口 Haikou	686.0	0.0	613.4
三 亚 Sanya	46.1	0.0	40.8

注：1. 表中数据为沿海地带合计数（表8-9、表8-13同）。
2. 沿海地带是指有海岸线的县、县级市、区（包括直辖市和地级市的区）。

Note: 1. The data in the table are the total numbers for the coastal regions (The same for Tables 8-9, 8-13).
2.Coastal Zone refers to the counties, county-level cities and districts with coastlines (including the districts under the municipalities directly under the Central Government and the prefecture-level districts)

8-7 沿海地区一般工业固体废物倾倒丢弃、处理及综合利用情况
Discharge, Treatment and Multipurpose Utilization of Common Industrial Solid Wastes by Coastal Regions

单位：万吨 (10 000 t)

地区 Region		一般工业固体废物倾倒丢弃量 Volume of Common Industrial Solid Wastes Discharged	一般工业固体废物处置量 Volume of Common Industrial Solid Wastes Treated	一般工业固体废物综合利用量 Volume of Common Industrial Solid Wastes Comprehensively Utilized
总计 Total		**8.6**	**36 176.3**	**80 310.8**
环渤海经济区 Round-the-Bohai Sea Economic Zone	**合计 Total**	**5.9**	**32 941.0**	**49 051.1**
	辽宁 Liaoning	5.9	9 421.7	10 719.2
	河北 Hebei	0.0	22 926.9	18 227.7
	天津 Tianjin	0.0	10.6	1 723.9
	山东 Shandong	0.0	581.7	18 380.2
长江三角洲经济区 Yangtze River Delta Economic Zone	**合计 Total**	**0.3**	**530.8**	**16 757.4**
	江苏 Jiangsu	0.3	278.9	10 577.8
	上海 Shanghai	0.0	47.0	1 876.9
	浙江 Zhejiang	0.0	204.9	4 302.8
海峡西岸经济区 Economic Zone on the West Side of the Taiwan Straits	**合计 Total**	**0.0**	**585.1**	**4 277.7**
	福建 Fujian	0.0	585.1	4 277.7
珠江三角洲经济区 Zhujiang River Delta Economic Zone	**合计 Total**	**1.9**	**630.6**	**4 893.0**
	广东 Guangdong	1.9	630.6	4 893.0
环北部湾经济区 Round-the-Beibu Gulf Economic Zone	**合计 Total**	**0.4**	**1 488.8**	**5 331.6**
	广西 Guangxi	0.4	1 454.4	5 057.7
	海南 Hainan	0.0	34.5	273.9

8-8 沿海城市一般工业固体废物倾倒丢弃、处理及综合利用情况
Discharge, Treatment and Multipurposed Utilization of Common Industrial Solid Wastes by Coastal Cities

单位：万吨 (10 000 t)

沿海城市 Coastal City	一般工业固体废物倾倒丢弃量 Volume of Common Industrial Solid Wastes Discharged	一般工业固体废物处置量 Volume of Common Industrial Solid Wastes Treated	一般工业固体废物综合利用量 Volume of Common Industrial Solid Wastes Comprehensively Utilized
天　津 Tianjin	0.0	10.6	1 723.9
唐　山 Tangshan	0.0	2 849.7	7 101.9
秦皇岛 Qinhuangdao	0.0	631.4	1 524.0
沧　州 Cangzhou	0.0	0.5	446.3
大　连 Dalian	0.0	89.5	458.2
丹　东 Dandong	0.0	300.7	274.2
锦　州 Jinzhou	0.0	17.4	235.8
营　口 Yingkou	0.0	0.8	655.6
盘　锦 Panjin	0.0	12.5	154.0
葫芦岛 Huludao	0.0	4.5	470.1
上　海 Shanghai	0.0	47.0	1 876.9
南　通 Nantong	0.0	7.9	499.1
连云港 Lianyungang	0.0	12.4	593.2
盐　城 Yancheng	0.0	17.9	273.9

8-8 续表1 continued

沿海城市 Coastal City	一般工业固体废物倾倒丢弃量 Volume of Common Industrial Solid Wastes Discharged	一般工业固体废物处置量 Volume of Common Industrial Solid Wastes Treated	一般工业固体废物综合利用量 Volume of Common Industrial Solid Wastes Comprehensively Utilized
杭　州 Hangzhou	0.0	61.1	656.2
宁　波 Ningbo	0.0	56.8	1 114.2
温　州 Wenzhou	0.0	4.8	268.7
嘉　兴 Jiaxing	0.0	20.6	499.1
绍　兴 Shaoxing	0.0	8.5	349.6
舟　山 Zhoushan	0.0	0.4	94.3
台　州 Taizhou	0.0	11.9	251.1
福　州 Fuzhou	0.0	31.2	750.6
厦　门 Xiamen	0.0	8.0	101.1
莆　田 Putian	0.0	5.4	67.5
泉　州 Quanzhou	0.0	6.3	741.8
漳　州 Zhangzhou	0.0	4.6	296.6
宁　德 Ningde	0.0	12.0	278.0
青　岛 Qingdao	0.0	28.2	842.4
东　营 Dongying	0.0	3.1	341.5
烟　台 Yantai	0.0	318.7	2 144.2
潍　坊 Weifang	0.0	21.1	823.4
威　海 Weihai	0.0	19.5	304.9
日　照 Rizhao	0.0	3.8	840.9
滨　州 Binzhou	0.0	27.2	1 638.6

8-8 续表2 continued

沿海城市 Coastal City	一般工业固体废物倾倒丢弃量 Volume of Common Industrial Solid Wastes Discharged	一般工业固体废物处置量 Volume of Common Industrial Solid Wastes Treated	一般工业固体废物综合利用量 Volume of Common Industrial Solid Wastes Comprehensively Utilized
广　州 Guangzhou	0.0	24.7	468.5
深　圳 Shenzhen	0.0	47.4	65.2
珠　海 Zhuhai	0.0	14.5	270.1
汕　头 Shantou	0.0	0.2	143.1
江　门 Jiangmen	0.0	25.3	226.2
湛　江 Zhanjiang	0.1	3.6	215.0
茂　名 Maoming	0.0	36.2	195.2
惠　州 Huizhou	0.0	3.0	91.0
汕　尾 Shanwei	0.0	0.5	60.7
阳　江 Yangjiang	0.2	0.1	330.6
东　莞 Dongguan	0.3	90.9	461.2
中　山 Zhongshan	0.6	26.3	73.2
潮　州 Chaozhou	0.0	0.1	75.0
揭　阳 Jieyang	0.0	0.1	155.5
北　海 Beihai	0.0	0.0	300.8
防城港 Fangchenggang	0.0	0.0	242.8
钦　州 Qinzhou	0.0	1.6	102.2
海　口 Haikou	0.0	1.2	4.0
三　亚 Sanya	0.0	0.1	0.0

8-9 沿海地带一般工业固体废物倾倒丢弃、处理及综合利用情况
Discharge, Treatment and Multipurposed Utilization of Common Industrial Solid Wastes by Coastal Counties

单位：万吨 (10 000 t)

地 区 Region	一般工业固体废物倾倒丢弃量 Volume of Common Industrial Solid Wastes Discharged	一般工业固体废物处置量 Volume of Common Industrial Solid Wastes Treated	一般工业固体废物综合利用量 Volume of Common Industrial Solid Wastes Comprehensively Utilized
天 津 Tianjin	**0.0**	**8.4**	**403.0**
河 北 Hebei	**0.0**	**105.2**	**2 151.3**
唐 山 Tangshan	0.0	59.9	1 010.7
秦皇岛 Qinhuangdao	0.0	45.3	1 138.3
沧 州 Cangzhou	0.0	0.0	2.3
辽 宁 Liaoning	**0.0**	**89.7**	**1 296.0**
大 连 Dalian	0.0	84.4	382.1
丹 东 Dandong	0.0	0.0	40.9
锦 州 Jinzhou	0.0	1.3	26.3
营 口 Yingkou	0.0	0.6	607.0
盘 锦 Panjin	0.0	0.0	17.9
葫芦岛 Huludao	0.0	3.4	222.0
上 海 Shanghai	**0.0**	**27.4**	**1 689.2**
江 苏 Jiangsu	**0.0**	**29.8**	**893.8**
南 通 Nantong	0.0	4.4	217.3
连云港 Lianyungang	0.0	9.7	472.4
盐 城 Yancheng	0.0	15.7	204.2

8-9 续表1 continued

地　区 Region	一般工业固体废物倾倒丢弃量 Volume of Common Industrial Solid Wastes Discharged	一般工业固体废物处置量 Volume of Common Industrial Solid Wastes Treated	一般工业固体废物综合利用量 Volume of Common Industrial Solid Wastes Comprehensively Utilized
浙　江　Zhejiang	**0.0**	**94.6**	**2 161.8**
杭　州　Hangzhou	0.0	18.2	135.2
宁　波　Ningbo	0.0	55.4	1 082.1
温　州　Wenzhou	0.0	3.1	233.9
嘉　兴　Jiaxing	0.0	1.8	223.7
绍　兴　Shaoxing	0.0	4.5	159.5
舟　山　Zhoushan	0.0	0.4	94.3
台　州　Taizhou	0.0	11.2	233.2
福　建　Fujian	**0.0**	**46.1**	**1 562.6**
福　州　Fuzhou	0.0	30.2	642.5
厦　门　Xiamen	0.0	8.0	101.1
莆　田　Putian	0.0	0.1	40.5
泉　州　Quanzhou	0.0	5.4	312.6
漳　州　Zhangzhou	0.0	1.9	201.2
宁　德　Ningde	0.0	0.5	264.7
山　东　Shandong	**0.0**	**371.6**	**4 352.1**
青　岛　Qingdao	0.0	26.9	639.5
东　营　Dongying	0.0	2.9	285.7
烟　台　Yantai	0.0	317.9	2 078.7
潍　坊　Weifang	0.0	1.8	273.7
威　海　Weihai	0.0	19.1	135.7
日　照　Rizhao	0.0	1.9	668.0
滨　州　Binzhou	0.0	1.1	270.7

8-9 续表2 continued

地 区 Region	一般工业固体废物倾倒丢弃量 Volume of Common Industrial Solid Wastes Discharged	一般工业固体废物处置量 Volume of Common Industrial Solid Wastes Treated	一般工业固体废物综合利用量 Volume of Common Industrial Solid Wastes Comprehensively Utilized
广 东 Guangdong	**1.1**	**247.8**	**2 151.6**
广 州 Guangzhou	0.0	20.2	391.5
深 圳 Shenzhen	0.0	45.9	64.6
珠 海 Zhuhai	0.0	2.3	9.8
汕 头 Shantou	0.0	0.2	143.1
江 门 Jiangmen	0.0	19.7	199.5
湛 江 Zhanjiang	0.0	3.5	211.4
茂 名 Maoming	0.0	36.1	97.4
惠 州 Huizhou	0.0	2.2	52.0
汕 尾 Shanwei	0.0	0.5	60.5
阳 江 Yangjiang	0.2	0.0	204.2
东 莞 Dongguan	0.3	90.9	461.2
中 山 Zhongshan	0.6	26.3	73.2
潮 州 Chaozhou	0.0	0.1	71.6
揭 阳 Jieyang	0.0	0.1	111.6
广 西 Guangxi	**0.0**	**0.6**	**533.2**
北 海 Beihai	0.0	0.0	300.8
防城港 Fangchenggang	0.0	0.0	168.9
钦 州 Qinzhou	0.0	0.6	63.5
海 南 Hainan	**0.0**	**1.3**	**2.4**
海 口 Haikou	0.0	1.2	2.4
三 亚 Sanya	0.0	0.1	0.0

8-10 主要沿海城市工业废气排放及处理情况
Emission and Treatment of Industrial Waste Gas in Major Coastal Cities

城市 City	工业废气排放量（亿立方米）Industrial Waste Gas Emission ($100\ 000\ 000\ m^3$)	工业二氧化硫排放量（吨）Industrial Sulphur Dioxide Emission (t)	工业二氧化硫去除量（吨）Industrial Sulphur Dioxide Treated (t)	工业氮氧化物排放量（吨）Industrial Nitrogen Oxide Emission (t)	工业氮氧化物处理量（吨）Industrial Nitrogen Oxide Treated (t)
天 津 Tianjin	8 800.0	195 395.1	654 737.0	216 946.7	150 843.3
唐 山 Tangshan	31 137.4	250 760.7	489 784.6	248 025.0	54 013.5
秦皇岛 Qinhuangdao	3 872.8	65 512.1	80 255.5	53 779.2	12 422.1
沧 州 Cangzhou	2 676.0	39 802.9	107 467.3	40 481.8	20 611.7
大 连 Dalian	4 686.8	94 370.3	149 686.1	94 647.6	20 731.9
丹 东 Dandong	1 257.9	30 543.8	49 027.6	18 925.8	9 903.0
锦 州 Jinzhou	1 156.0	38 535.2	45 147.2	20 189.2	3 269.6
营 口 Yingkou	4 293.2	56 091.5	53 106.0	48 189.9	8 682.6
盘 锦 Panjin	996.1	55 450.2	47 897.3	27 155.1	729.3
葫芦岛 Huludao	1 398.7	58 190.6	495 126.3	38 083.1	29 537.8
上 海 Shanghai	13 007.4	155 360.2	323 505.2	228 621.3	50 458.6
南 通 Nantong	2 412.8	61 811.6	183 039.6	50 779.2	11 211.9
连云港 Lianyungang	2 493.8	47 568.9	67 095.4	27 986.2	7 106.9
盐 城 Yancheng	1 364.5	45 518.7	86 975.1	32 448.7	12 348.9
杭 州 Hangzhou	4 428.3	80 349.4	72 026.0	61 626.8	12 455.5
宁 波 Ningbo	6 365.6	118 101.6	405 469.2	162 112.7	69 089.4
温 州 Wenzhou	4 059.4	34 125.1	94 796.0	35 628.7	14 902.2

8-10 续表1 continued

城 市 City	工业废气排放量（亿立方米）Industrial Waste Gas Emission (100 000 000 m^3)	工业二氧化硫排放量（吨）Industrial Sulphur Dioxide Emission (t)	工业二氧化硫去除量（吨）Industrial Sulphur Dioxide Treated (t)	工业氮氧化物排放量（吨）Industrial Nitrogen Oxide Emission (t)	工业氮氧化物处理量（吨）Industrial Nitrogen Oxide Treated (t)
嘉 兴 Jiaxing	2 324.0	77 133.1	118 425.1	43 383.2	21 221.5
绍 兴 Shaoxing	1 726.7	64 935.3	48 621.5	41 944.3	6 771.1
舟 山 Zhoushan	357.5	11 481.7	27 996.8	13 705.3	3 705.5
台 州 Taizhou	1 313.1	28 083.4	129 388.9	31 461.9	4 543.6
福 州 Fuzhou	3 661.7	56 385.0	174 635.3	71 392.1	47 941.8
厦 门 Xiamen	1 512.5	16 144.1	37 912.6	9 350.3	24 111.4
莆 田 Putian	584.6	9 075.5	22 064.1	13 653.7	12 315.8
泉 州 Quanzhou	3 631.6	110 699.4	158 479.9	67 672.0	26 433.5
漳 州 Zhangzhou	1 829.2	37 649.7	112 287.4	52 950.7	10 009.2
宁 德 Ningde	1 116.0	16 706.5	47 779.9	9 961.2	34 226.4
青 岛 Qingdao	2 084.9	64 029.2	217 848.3	58 486.7	56 551.6
东 营 Dongying	1 160.7	48 312.0	379 495.4	28 693.3	21 521.6
烟 台 Yantai	3 223.8	75 463.8	370 944.0	75 945.4	19 021.9
潍 坊 Weifang	3 085.5	120 567.4	254 820.1	63 059.3	18 789.0
威 海 Weihai	940.5	30 669.1	110 928.7	26 864.7	8 767.8
日 照 Rizhao	5 152.7	41 228.3	137 699.5	47 916.1	15 539.5
滨 州 Binzhou	4 193.6	140 026.4	891 152.1	143 680.6	41 122.3
广 州 Guangzhou	3 715.2	61 059.4	381 457.0	51 606.6	48 013.9

8-10 续表2 continued

城 市 City	工业废气排放量（亿立方米） Industrial Waste Gas Emission (100 000 000 m^3)	工业二氧化硫排放量（吨） Industrial Sulphur Dioxide Emission (t)	工业二氧化硫去除量（吨） Industrial Sulphur Dioxide Treated (t)	工业氮氧化物排放量（吨） Industrial Nitrogen Oxide Emission (t)	工业氮氧化物处理量（吨） Industrial Nitrogen Oxide Treated (t)
深 圳 Shenzhen	1 847.5	5 339.5	38 565.9	14 348.7	15 701.8
珠 海 Zhuhai	1 693.5	20 680.9	45 683.0	43 094.8	230.3
汕 头 Shantou	871.8	26 784.0	72 790.9	19 318.2	8 431.7
江 门 Jiangmen	1 491.2	51 691.8	123 242.1	50 617.6	27 999.5
湛 江 Zhanjiang	1 002.2	20 376.8	82 721.1	11 978.9	12 647.5
茂 名 Maoming	1 001.2	29 177.4	466 725.2	16 263.1	5 238.2
惠 州 Huizhou	1 606.4	28 941.0	36 820.3	41 628.2	13 642.8
汕 尾 Shanwei	419.2	10 427.4	53 722.1	9 444.9	6 383.6
阳 江 Yangjiang	1 098.1	24 128.9	51 477.9	22 262.3	13 262.1
东 莞 Dongguan	3 175.2	115 316.6	174 894.0	102 977.1	22 341.8
中 山 Zhongshan	874.2	22 277.7	9 892.4	21 588.8	414.4
潮 州 Chaozhou	780.3	12 578.3	59 400.7	12 182.6	10 183.2
揭 阳 Jieyang	876.9	23 828.8	79 689.5	9 379.6	9 780.6
北 海 Beihai	656.5	11 688.5	26 848.4	6 293.4	9 917.0
防城港 Fangchenggang	1 075.4	24 444.7	46 290.2	14 869.2	6 431.2
钦 州 Qinzhou	563.5	16 313.7	31 290.3	7 289.5	4 889.5
海 口 Haikou	64.3	1 772.8	0.0	172.5	0.0
三 亚 Sanya	58.7	2.4	0.0	178.1	32.8

8-10 续表3 continued

城 市 City	工业烟（粉）尘排放量（亿立方米） Industrial Soot (Dust) Emission (100 000 000 m^3)	工业烟粉尘处理量（吨） Industrial Soot (Dust) Treated (t)	生活二氧化硫排放量（吨） Household Sulphur Dioxide Emission (t)	生活氮氧化物排放量（吨） Household Nitrogen Oxide Emission (t)	生活烟尘排放量（吨） Household Soot Emission (t)
天 津 Tianjin	112 187.0	4 802 198.4	13 767.2	9 516.7	21 072.1
唐 山 Tangshan	536 091.7	15 742 681.8	21 446.2	10 710.0	54 000.0
秦皇岛 Qinhuangdao	59 220.9	2 208 722.4	8 030.5	1 668.3	17 855.6
沧 州 Cangzhou	63 450.5	1 468 894.5	8 553.9	1 107.6	17 860.0
大 连 Dalian	72 465.0	3 735 417.4	16 365.9	2 403.0	11 666.4
丹 东 Dandong	45 562.6	1 212 350.0	4 194.6	1 099.1	7 035.8
锦 州 Jinzhou	47 705.9	789 932.7	5 116.1	928.8	7 410.0
营 口 Yingkou	96 395.6	2 484 329.1	3 857.5	774.9	14 787.7
盘 锦 Panjin	15 830.4	893 798.5	3 876.0	618.3	2 588.8
葫芦岛 Huludao	37 679.3	1 427 824.9	2 672.6	414.1	5 917.5
上 海 Shanghai	131 432.5	5 926 060.9	32 764.9	18 734.2	4 017.1
南 通 Nantong	41 362.4	2 882 921.1	3 060.0	332.0	1 200.0
连云港 Lianyungang	41 367.7	1 503 589.6	5 960.2	786.8	2 874.9
盐 城 Yancheng	52 531.6	1 332 911.0	2 317.2	351.0	1 281.1
杭 州 Hangzhou	70 346.5	4 713 833.5	632.8	335.4	135.5
宁 波 Ningbo	30 577.4	6 566 998.0	1 876.8	480.0	3 000.0
温 州 Wenzhou	18 390.3	1 330 365.2	668.6	423.3	467.5

8-10 续表4 continued

城 市 City	工业烟（粉）尘排放量（亿立方米） Industrial Soot (Dust) Emission (100 000 000 m^3)	工业烟粉尘处理量（吨） Industrial Soot (Dust) Treated (t)	生活二氧化硫排放量（吨） Household Sulphur Dioxide Emission (t)	生活氮氧化物排放量（吨） Household Nitrogen Oxide Emission (t)	生活烟尘排放量（吨） Household Soot Emission (t)
嘉 兴 Jiaxing	30 127.3	2 893 507.4	385.0	93.0	45.0
绍 兴 Shaoxing	36 206.0	1 693 782.6	734.4	201.6	270.0
舟 山 Zhoushan	5 006.3	320 469.0	288.0	98.3	180.6
台 州 Taizhou	13 106.0	950 729.6	788.6	136.4	93.7
福 州 Fuzhou	105 712.9	2 780 658.6	1 279.0	169.0	547.0
厦 门 Xiamen	4 561.0	419 302.4	180.9	65.6	85.1
莆 田 Putian	5 232.1	217 947.4	1 406.2	451.7	1 054.7
泉 州 Quanzhou	68 355.5	1 913 256.4	3 195.2	557.7	1 769.0
漳 州 Zhangzhou	24 684.0	1 188 315.7	795.3	117.0	467.8
宁 德 Ningde	13 524.1	797 131.1	846.3	70.0	534.4
青 岛 Qingdao	32 196.1	2 372 606.8	27 088.6	2 647.9	9 805.8
东 营 Dongying	7 194.3	1 416 094.5	1 720.9	864.7	925.5
烟 台 Yantai	34 690.8	4 067 338.2	9 837.0	2 058.8	6 651.4
潍 坊 Weifang	63 411.0	5 323 746.6	12 732.9	1 650.6	10 206.7
威 海 Weihai	12 312.9	1 297 230.2	11 908.3	1 352.9	2 956.2
日 照 Rizhao	98 860.7	3 360 227.3	16 633.0	3 255.3	14 900.0
滨 州 Binzhou	94 304.2	8 582 795.7	3 642.6	555.6	2 610.9
广 州 Guangzhou	22 135.9	3 102 457.1	2 363.0	720.0	214.0

8-10 续表5 continued

城市 City	工业烟（粉）尘排放量（亿立方米） Industrial Soot (Dust) Emission (100 000 000 m^3)	工业烟粉尘处理量（吨） Industrial Soot (Dust) Treated (t)	生活二氧化硫排放量（吨） Household Sulphur Dioxide Emission (t)	生活氮氧化物排放量（吨） Household Nitrogen Oxide Emission (t)	生活烟尘排放量（吨） Household Soot Emission (t)
深圳 Shenzhen	2 677.5	411 002.2	238.0	885.0	82.0
珠海 Zhuhai	12 971.5	512 546.8	0.0	3.3	0.0
汕头 Shantou	8 145.2	695 169.6	86.2	117.0	55.2
江门 Jiangmen	17 723.2	1 370 894.7	110.0	4.2	70.0
湛江 Zhanjiang	11 246.9	815 503.0	2 611.2	407.7	266.2
茂名 Maoming	13 510.9	730 252.7	2 487.1	596.6	2 016.6
惠州 Huizhou	24 648.0	2 186 364.6	258.1	86.0	67.3
汕尾 Shanwei	4 226.2	413 020.8	616.4	82.9	477.8
阳江 Yangjiang	31 646.6	897 862.2	1 275.0	162.3	400.0
东莞 Dongguan	17 892.6	1 603 113.9	1 561.0	156.0	450.0
中山 Zhongshan	16 703.0	233 436.0	238.0	45.8	200.0
潮州 Chaozhou	4 510.8	574 034.8	1 035.0	152.2	114.2
揭阳 Jieyang	8 990.2	615 398.7	4 318.0	635.0	2 540.0
北海 Beihai	3 312.6	368 789.6	1 282.0	145.4	408.5
防城港 Fangchenggang	49 786.7	1 331 315.8	960.0	100.0	400.0
钦州 Qinzhou	4 670.0	215 948.0	1 117.0	76.7	237.3
海口 Haikou	998.4	2 552.1	15.0	18.5	229.3
三亚 Sanya	1 188.1	20 129.9	69.9	12.2	6.2

8-11 沿海地区污染治理项目情况
Pollution Treatment Projects in Coastal Regions

单位：个 (unit)

地区 Region		当年安排施工项目 Arranged for Construction in the Year		当年竣工项目 Completed in the Year	
		治理废水 Treatment of Waste Water	治理固体废物 Treatment of Solid Wastes	治理废水 Treatment of Waste Water	治理固体废物 Treatment of Solid Wastes
总 计 Total		**784**	**117**	**973**	**116**
环渤海经济区 Round-the-Bohai Sea Economic Zone	**合 计 Total**	**116**	**18**	**139**	**18**
	辽 宁 Liaoning	12	5	9	5
	河 北 Hebei	21	2	36	2
	天 津 Tianjin	17	0	17	0
	山 东 Shandong	66	11	77	11
长江三角洲经济区 Yangtze River Delta Economic Zone	**合 计 Total**	**421**	**26**	**562**	**30**
	江 苏 Jiangsu	136	4	186	3
	上 海 Shanghai	39	1	38	1
	浙 江 Zhejiang	246	21	338	26
海峡西岸经济区 Economic Zone on the West Side of the Taiwan Straits	**合 计 Total**	**69**	**62**	**54**	**59**
	福 建 Fujian	69	62	54	59
珠江三角洲经济区 Zhujiang River Delta Economic Zone	**合 计 Total**	**130**	**5**	**168**	**4**
	广 东 Guangdong	130	5	168	4
环北部湾经济区 Round-the-Beibu Gulf Economic Zone	**合 计 Total**	**48**	**6**	**50**	**5**
	广 西 Guangxi	48	6	49	5
	海 南 Hainan	0	0	1	0

8-12 沿海城市污染治理项目情况
Pollution Treatment Projects in Coastal Cities

单位：个 (unit)

沿海城市 Coastal City	当年安排施工项目 Arranged for Construction in the Year		当年竣工项目 Completed in the Year	
	治理废水 Treatment of Waste Water	治理固体废物 Treatment of Solid Wastes	治理废水 Treament of Waste Water	治理固体废物 Treatment of Solid Wastes
天 津 Tianjin	17	0	17	0
唐 山 Tangshan	8	1	14	1
秦皇岛 Qinhuangdao	0	0	7	0
沧 州 Cangzhou	1	0	1	0
大 连 Dalian	1	0	0	0
丹 东 Dandong	0	0	0	0
锦 州 Jinzhou	0	2	0	0
营 口 Yingkou	0	0	0	0
盘 锦 Panjin	0	2	0	4
葫芦岛 Huludao	3	0	2	0
上 海 Shanghai	39	1	38	1
南 通 Nantong	28	0	37	1
连云港 Lianyungang	20	0	14	0
盐 城 Yancheng	1	0	1	0

8-12 续表1 continued

沿海城市 Coastal City	当年安排施工项目 Arranged for Construction in the Year		当年竣工项目 Completed in the Year	
	治理废水 Treatment of Waste Water	治理固体废物 Treatment of Solid Wastes	治理废水 Treament of Waste Water	治理固体废物 Treatment of Solid Wastes
杭 州 Hangzhou	29	5	45	5
宁 波 Ningbo	11	1	9	0
温 州 Wenzhou	13	0	12	0
嘉 兴 Jiaxing	41	1	51	1
绍 兴 Shaoxing	43	2	86	7
舟 山 Zhoushan	1	0	1	0
台 州 Taizhou	27	2	21	2
福 州 Fuzhou	3	1	1	1
厦 门 Xiamen	11	1	16	1
莆 田 Putian	0	1	0	0
泉 州 Quanzhou	31	58	12	56
漳 州 Zhangzhou	5	0	8	0
宁 德 Ningde	1	0	2	0
青 岛 Qingdao	2	0	2	0
东 营 Dongying	1	0	2	0
烟 台 Yantai	8	1	7	1
潍 坊 Weifang	20	2	18	1
威 海 Weihai	2	0	1	0
日 照 Rizhao	1	0	1	0
滨 州 Binzhou	0	0	0	0

8-12 续表2 continued

沿海城市 Coastal City	当年安排施工项目 Arranged for Construction in the Year		当年竣工项目 Completed in the Year	
	治理废水 Treatment of Waste Water	治理固体废物 Treatment of Solid Wastes	治理废水 Treament of Waste Water	治理固体废物 Treatment of Solid Wastes
广　州　Guangzhou	7	0	11	0
深　圳　Shenzhen	9	0	10	0
珠　海　Zhuhai	11	0	12	0
汕　头　Shantou	3	0	13	0
江　门　Jiangmen	25	1	28	1
湛　江　Zhanjiang	12	1	14	0
茂　名　Maoming	0	0	2	0
惠　州　Huizhou	12	0	12	0
汕　尾　Shanwei	0	0	0	0
阳　江　Yangjiang	3	0	5	0
东　莞　Dongguan	9	0	17	1
中　山　Zhongshan	9	0	10	0
潮　州　Chaozhou	1	0	2	0
揭　阳　Jieyang	0	0	0	0
北　海　Beihai	5	0	4	0
防城港　Fangchenggang	0	0	0	0
钦　州　Qinzhou	1	0	2	0
海　口　Haikou	0	0	0	0
三　亚　Sanya	0	0	0	0

8-13　沿海地带污染治理项目情况
Pollution Treatment Projects in Coastal Counties

单位：个　　(unit)

地　区 Region	当年安排施工项目 Arranged for Construction in the Year		当年竣工项目 Completed in the Year	
	治理废水 Treatment of Waste Water	治理固体废物 Treatment of Solid Wastes	治理废水 Treament of Waste Water	治理固体废物 Treatment of Solid Wastes
天　津　Tianjin	**9**	**0**	**9**	**0**
河　北　Hebei	**2**	**0**	**3**	**0**
唐　山　Tangshan	2	0	2	0
秦皇岛　Qinhuangdao	0	0	1	0
沧　州　Cangzhou	0	0	0	0
辽　宁　Liaoning	**4**	**0**	**2**	**0**
大　连　Dalian	1	0	0	0
丹　东　Dandong	0	0	0	0
锦　州　Jinzhou	0	0	0	0
营　口　Yingkou	0	0	0	0
盘　锦　Panjin	0	0	0	0
葫芦岛　Huludao	3	0	2	0
上　海　Shanghai	**31**	**1**	**28**	**1**
江　苏　Jiangsu	**42**	**0**	**44**	**1**
南　通　Nantong	22	0	29	1
连云港　Lianyungang	19	0	14	0
盐　城　Yancheng	1	0	1	0

8-13 续表1 continued

地 区 Region	当年安排施工项目 Arranged for Construction in the Year		当年竣工项目 Completed in the Year	
	治理废水 Treatment of Waste Water	治理固体废物 Treatment of Solid Wastes	治理废水 Treament of Waste Water	治理固体废物 Treatment of Solid Wastes
浙 江 Zhejiang	**106**	**9**	**115**	**9**
杭 州 Hangzhou	18	4	24	5
宁 波 Ningbo	11	1	9	0
温 州 Wenzhou	12	0	11	0
嘉 兴 Jiaxing	14	1	16	1
绍 兴 Shaoxing	27	1	35	1
舟 山 Zhoushan	1	0	1	0
台 州 Taizhou	23	2	19	2
福 建 Fujian	**41**	**9**	**28**	**7**
福 州 Fuzhou	2	1	1	1
厦 门 Xiamen	11	1	16	1
莆 田 Putian	0	1	0	0
泉 州 Quanzhou	25	6	5	5
漳 州 Zhangzhou	2	0	4	0
宁 德 Ningde	1	0	2	0
山 东 Shandong	**23**	**2**	**19**	**2**
青 岛 Qingdao	0	0	0	0
东 营 Dongying	1	0	2	0
烟 台 Yantai	7	1	6	1
潍 坊 Weifang	12	1	9	1
威 海 Weihai	2	0	1	0
日 照 Rizhao	1	0	1	0
滨 州 Binzhou	0	0	0	0

8-13 续表2 continued

地 区 Region	当年安排施工项目 Arranged for Construction in the Year		当年竣工项目 Completed in the Year	
	治理废水 Treatment of Waste Water	治理固体废物 Treatment of Solid Wastes	治理废水 Treament of Waste Water	治理固体废物 Treatment of Solid Wastes
广 东 Guangdong	**78**	**1**	**111**	**1**
广 州 Guangzhou	5	0	10	0
深 圳 Shenzhen	9	0	9	0
珠 海 Zhuhai	5	0	6	0
汕 头 Shantou	3	0	13	0
江 门 Jiangmen	13	0	14	0
湛 江 Zhanjiang	12	1	14	0
茂 名 Maoming	0	0	2	0
惠 州 Huizhou	10	0	10	0
汕 尾 Shanwei	0	0	0	0
阳 江 Yangjiang	3	0	5	0
东 莞 Dongguan	9	0	17	1
中 山 Zhongshan	9	0	10	0
潮 州 Chaozhou	0	0	1	0
揭 阳 Jieyang	0	0	0	0
广 西 Guangxi	**6**	**0**	**5**	**0**
北 海 Beihai	5	0	4	0
防城港 Fangchenggang	0	0	0	0
钦 州 Qinzhou	1	0	1	0
海 南 Hainan	**0**	**0**	**0**	**0**
海 口 Haikou	0	0	0	0
三 亚 Sanya	0	0	0	0

8-14 沿海区域海洋类型自然保护区建设情况

Construction of Marine-Type Nature Reserves in Coastal Regions

地 区 Region	保护区数量（个） Number of Nature Reserves (unit)	按保护级别分（个） By Level of Protection (unit)		按保护类型分（个） By Type of Protection (unit)				保护区面积（平方千米） Area(km²)
		国家级 National	地方级 Provincial	海洋和海岸生态系统 Marine and Coastal Ecosystem	海洋自然历史遗迹 Marine Natural and Historical Relics	海洋生物多样性 Marine Biodiversity	其他 Others	
合 计 Total	**147**	**32**	**115**	**69**	**7**	**65**	**6**	**48 455**
环渤海经济区 Round-the-Bohai Sea Economic Zone	50	15	35	26	3	21	0	16 319
长江三角洲经济区 Yangtze River Delta Economic Zone	14	3	11	3	1	6	4	2 391
海峡西岸经济区 Economic Zone on the West Side of the Taiwan Straits	12	3	9	5	2	5	0	1 089
珠江三角洲经济区 Zhujiang River Delta Economic Zone	50	5	45	20	0	28	2	3 820
环北部湾经济区 Round-the-Beibu Gulf Economic Zone	21	6	15	15	1	5	0	24 837

8-15 沿海地区海洋类型自然保护区建设情况
Construction of Marine-Type Nature Reserves

地　区 Region	保护区数量（个） Number of Nature Reserves (unit)	按保护级别分（个） By Level of Protection (unit)		按保护类型分（个） By Type of Protection (unit)				保护区面积（平方千米） Area (km^2)
		国家级 National	地方级 Provincial	海洋和海岸生态系统 Marine and Coastal Ecosystem	海洋自然历史遗迹 Marine Natural and Historical Relics	海洋生物多样性 Marine Biodiversity	其他 Others	
合　计 Total	**147**	**32**	**115**	**69**	**7**	**65**	**6**	**48 455**
天　津 Tianjin	1	1	0	1	0	0	0	359
河　北 Hebei	3	1	2	2	1	0	0	344
辽　宁 Liaoning	15	5	10	10	2	3	0	9 860
上　海 Shanghai	4	0	4	1	0	0	3	941
江　苏 Jiangsu	4	1	3	0	0	4	0	724
浙　江 Zhejiang	6	2	4	2	1	2	1	726
福　建 Fujian	12	3	9	5	2	5	0	1 089
山　东 Shandong	31	8	23	13	0	18	0	5 755
广　东 Guangdong	50	5	45	20	0	28	2	3 820
广　西 Guangxi	2	2	0	2	0	0	0	110
海　南 Hainan	19	4	15	13	1	5	0	24 727

8-16 全国海洋生态监控区基本情况
Basic Condition of the Marine Ecological Monitoring Areas throughout the Country

生态监控区 Ecological Monitoring Area	所在地 Location	面积 （平方千米） Area (km^2)	主要生态系统类型 Major Types of Ecosystem	多样性指数* Diversity Indices		
				浮游植物 Phyto-plankton	大型浮游动物 Macrozoo-plankton	底栖生物 Macrobenthos
双台子河口 Shungtaizi Estuary	辽宁省 Liaoning Province	3 000	河　口 Estuary	2. 6	1. 8	
锦州湾 Jinzhou Bay	辽宁省 Liaoning Province	650	海　湾 Bay	3. 0	1. 1	
滦河口-北戴河 Luanhe River Mouth-Beidaihe	河北省 Hebei Province	900	河　口 Estuary	2. 8	1. 8	1. 2
渤海湾 Bohai Bay	天津市 Tianjin Municipality	3 000	海　湾 Bay	2. 1	2. 4	2. 9
莱州湾 Laizhou Bay	山东省 Shandong Province	3 770	海　湾 Bay	2. 5	2. 0	3. 2
黄河口 Yellow River Mouth	山东省 Shandong Province	2 600	河　口 Estuary	2. 6	0. 8	2. 2
苏北浅滩 North Jiangsu Bank	江苏省 Jiangsu Province	15 400	滩涂湿地 Mudflat and Wetland	2. 2	1. 6	1. 6
长江口 Yangtze River Mouth	上海市 Shanghai Municipality	13 668	河　口 Estuary	1. 7	2. 1	2. 5
杭州湾 Hangzhou Bay	上海市 浙江省 Shanghai Municipality Zhejiang Province	5 000	海　湾 Bay	2. 1	1. 6	0. 1
乐清湾 Yueqing Bay	浙江省 Zhejiang Province	464	海　湾 Bay	2. 9	2. 1	0. 9

8-16 续表 continued

生态监控区 Ecological Monitoring Area	所在地 Location	面积（平方千米） Area (km^2)	主要生态系统类型 Major Types of Ecosystem	多样性指数* Diversity Indices		
				浮游植物 Phyto-plankton	大型浮游动物 Macrozoo-plankton	底栖生物 Macrobenthos
闽东沿岸 Coastal East Fujian	福建省 Fujian Province	5 063	海 湾 Bay	1.9	2.4	2.4
大亚湾 Daya Bay	广东省 Guangdong Province	1 200	海 湾 Bay	2.0	3.5	1.9
珠江口 Zhujiang River Mouth	广东省 Guangdong Province	3 980	河 口 Estuary	1.7	3.5	1.2
雷州半岛西南沿岸 Southwest Coast of Leizhou Peninsula	广东省 Guangdong Province	1 150	珊瑚礁 Coral Reef			
广西北海 Beihai, Guangxi	广西壮族自治区 Guangxi Zhuang Nationality Autonomous Region	120	珊瑚礁 红树林 海草床 Coral Reef Mangroves Seagrass Bed			
北仑河口 Beilun River Mouth	广西壮族自治区 Guangxi Zhuang Nationality Autonomous Region	150	红树林 Mangroves			
海南东海岸 East Coast of Hainan	海南省 Hainan Province	3 750	珊瑚礁 海草床 Coral Reef Seagrass Bed			
西沙珊瑚礁 Xisha Coral Reef	海南省 Hainan Province	400	珊瑚礁 Coral Reef			

注：*生物多样性指数：是生物种数和种类间个体数量分配均匀性的综合表现，用Shannon-Wiener多样性指数表征。

Note: Biodiversity index: refers to the comprehensive expression of the distributive homogeneity of the number of biological species and the number of individuals between varieties characterized by the Shannon-Wiener biodiversity index.

8-17 沿海地区风暴潮灾害情况
Survey of Storm Surges by Coastal Regions

受灾地区 Disaster Area	受灾人口（万人） Disaster-stricken Population (10 000 persons)	死亡人数*（人） Death Toll (person)	受灾面积 Disaster-stricken area 农田（千公顷） Disaster-Affected farmland (1 000 hm^2)	受灾面积 Disaster-stricken area 水产养殖（千公顷） Affected Area of Mariculture (1 000 hm^2)
合 计 Total	**1 095.54**		**30.2**	**106.7**
江 苏 Jiangsu				12.8
浙 江 Zhejiang	93.1			9.6
福 建 Fujian	8.7		2.5	14.2
山 东 Shandong			0.8	1.4
广 东 Guangdong	515.0		0.7	45.4
广 西 Guangxi	224.8		3.7	8.8
海 南 Hainan	254.0		22.5	14.6

8-17 续表 continued

受灾地区 Disaster Area	海岸工程（千米）Coastal Engineering (km)	房屋 House（间）(unit)	船只（艘）Boats (unit)	直接经济损失（亿元）Direct Economic Loss (100 million yuan)
合 计 Total	**179.9**	**34 573.0**	**6 394.0**	**135.8**
江 苏 Jiangsu	23.8			0.5
浙 江 Zhejiang	6.7	60.0	202.0	4.3
福 建 Fujian	13.7	53.0	413.0	4.3
山 东 Shandong	11.3	69.0	21.0	1.4
广 东 Guangdong	20.4	11 717.0	1 213.0	60.4
广 西 Guangxi	76.0		501.0	28.3
海 南 Hainan	28.1	22 674.0	4 044.0	36.6

注：*包括失踪人数。

Notes:*Includes lost people.

8-18 沿海地区赤潮灾害情况
Survey of Red Tide by Coastal Regions

时间 Date	影响区域 Area	最大面积 （平方千米） Disaster Area(km^2)
累 计 In the Aggregate		**6 588**
其中：Including:		
8月26日至9月25日 Aug.26-Sep.25	天津滨海旅游区附近海域 Sea Area near the Tianjian Coastal Tourist Zone	300
6月11日至6月15日 Jun.11-Jun.15	河北省秦皇岛近岸海域 Nearshore Sea Area of Hebei Province Qinhuangdao	228
9月13日至9月17日 Sep.13-Sep.17	河北省渤海中部海域 Central Area of the Bohai Sea, Hebei Province	400
5月15日至8月7日 May.15-Aug.7	河北省秦皇岛近岸海域 Nearshore Sar Area of Qinhuangdao, Hebei Province	2 000
5月30日至6月13日 May.30-Jun.13	辽宁省辽东湾东部海域 Eastern Sea Area of the Liaodong Bay, Liaoning Province	110
5月21日至6月5日 May.21-Jun.5	浙江省舟山嵊泗海域 Shengsi Sea Area of Zhongshan, Zhejiang Province	170
9月7日至9月9日 Sep.7-Sep.9	浙江省舟山嵊泗海域 Shengsi Sea Area of Zhoushan, Zhejiang Province	200
5月21日至6月3日 May.21-Jun.3	浙江省舟山普陀海域 Putuo Sea Area of Zhoushan, Zhejiang Province	300

8-18 续表 continued

时间 Date	影响区域 Area	最大面积（平方千米） Disaster Area(km^2)
5月27日至6月3日 May.27-Jun.3	浙江省舟山朱家尖海域 Zhujiajian Sea Area of Zhoushan, Zhejiang Province	400
5月21日至6月9日 May.21-Jun.9	浙江省台州温岭海域 Wenling Sea Area of Taizhou, Zhejiang Province	100
5月19日至6月11日 May.19-Jun.11	浙江省温州苍南海域 Cangnan Sea Area of Wenzhou, Zhejiang Province	320
5月8日至5月15日 May.8-May.15	福建省莆田南日岛附近海域 Sea Area near Nanri Island, Putian, Fujian Province	600
9月21日至9月23日 Sep.21-Sep.23	山东省烟台长岛县附近海域 Sea Area near Changdao County Yantai, Shandong Province	890
4月11日至4月23日 Apr.11-Apr.23	广东省惠州市大亚湾马鞭洲以北海域和澳头湾 Aotou Bay and Sea Area North of Mabianzhou in the Daya Bay, Huizhou City, Guangdong Province	100
11月25日至11月27日 Nov.25-Nov.27	广东省茂名市博贺港鸡岛附近海域 Sea Area near Jidao, Bohe Harbour, Maoming City, Guangdong Province	300
7月21日至8月13日 Jul.21-Aug.13	广东省湛江沙湾至乌石港渔业增养殖水域 Waters for Fisheries Culture and Stock Enhancement from Shawan to Wushigang, Zhanjiang, Guangdong Province	140

注：本表仅列出最大面积超过100平方千米（含）的赤潮过程。

Note: This table only lists the red tide process that the maximum area exceeds 100 km^2.

主要统计指标解释

1. 工业废水排放量 指经过企业厂区所有排放口排到企业外部的工业废水量。包括生产废水、外排的直接冷却水、超标排放的矿井地下水和与工业废水混排的厂区生活污水,不包括外排的间接冷却水(清污不分流的间接冷却水应计算在内)。

2. 直接排入海的工业废水量 指经企业位于海边的排放口, 直接排入海中的废水量。直接排入是指废水经过工厂的排污口直接排入海，而未经过城市下水道或其他中间体，也不受其他水体的影响。

3. 工业废水处理量 指报告期内各种水治理设施实际处理的工业废水量,包括处理后外排的和处理后回用的工业废水量，虽经处理但未达到国家或地方排放标准的废水量也应计算在内。计算时，如遇车间和厂排放口均有治理设施，并对同一废水分级处理时，不应重复计算工业废水处理量。

4. 一般工业固体废物处置量 指将固体废物焚烧或者最终置于符合环境保护规定要求的场所并不再回取的工业固体废物量(包括当年处置往年的工业固体废物累计贮存量)。

5. 一般工业固体废物倾倒丢弃量 指将所产生的固体废物倾倒丢弃到固体废物污染防治设施、场所以外的量。不包括矿山开采的剥离废石和掘进废石(煤矸石和呈酸性或碱性的废石除外)。

6. 当年安排施工项目数 指报告期内由国家、部门、地方或企业单位安排开工的，并以治理废水、废气、固体废物、噪声和其他（如电磁波、恶臭等）环境污染的环境治理工程的总数。不包括“三同时”项目。

7. 当年竣工项目数 指报告期内竣工投入运行的治理废水、废气、固体废物、噪声及其他污染的环境工程项目的总数。

Explanatory Notes on Main Statistical Indicators

1. Volume of Industrial Waste Water Discharged refers to the quantity of industrial waste water discharged externally through all the outlets in the factory area of the enterprise, including the waste water from production, externally discharged direct cooling water, mine-shaft groundwater discharged exceeding the set standard and the domestic sewage of the factory area discharged together with the industrial waste water, but not including the indirect cooling water discharge externally.

2. Volume of Industrial Waste Water discharged Directly to Sea refers to the quantity of waste water directly discharged into the sea through the outlets of the enterprise by the sea. Direct discharge means the direct discharge into the sea of waste water through the outlets of the factory, which is not discharged via the urban sewers or other intermediates and is not affected by other water bodies.

3. Volume of Industrial Waste Water Treated refers to the industrial waste water volume actually treated by various water treatment facilities in the period covered by the report, including the quantity of the industrial waste water discharged and reused after treatment. The amount of the waste water which is not up to the state or local standard of discharge upon treatment should be included. If the workshops and outlets of the factory are provided with treatment facilities and carry out graded treatment of the same waste water, the processed volume of industrial waste water cannot be calculated repeatedly.

4. Volume of Common Industrial Solid Wastes Discharged refers to the amount of solid wastes discharged out of the facilities and sites for the solid wastes pollution prevention and control, not including the stripped and tunneled waste ores in the excavation of mines (other than gangues and acid or alkaline waste ores).

5. Volume of Common Industrial Solid Wastes Treated refers to the volume of industrial solid wastes which are to be burned or finally placed at the sites in keeping with the requirement of environmental protection and will not be recovered (including the accumulated amount of storage in former years of industrial solid radioactive matter disposed of in the year).

6. Arranged for Construction in the Year refers to the total number of environmental pollution control projects for controlling waste water, waste gas, solid wastes, noise and other environmental pollutions (such as electromagnetic wave, offensive odor) started by the state, governmental departments, local governments or enterprises in the period covered by the report mainly for the purpose of pollution control and multipurpose utilization of "three wastes".

7. Number of Pollution Treatment Projects Completed in the Current Year refers to the total number of environmental engineering projects for controlling waste water, waste gas, solid wastes, noise and other pollutions which are completed and put into operation in the period covered by the report.

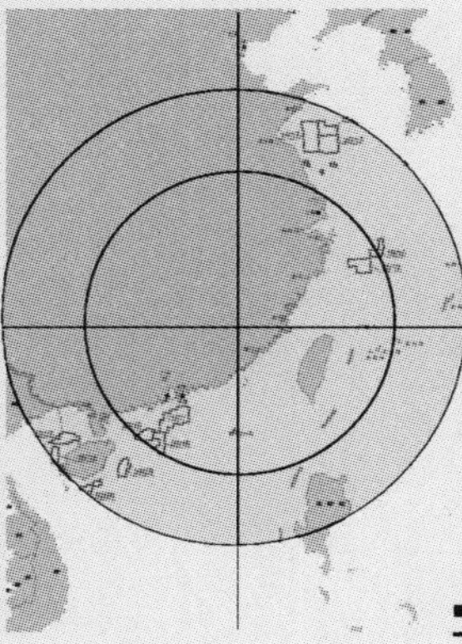

9 海洋行政管理及公益服务

Marine Administration and Public-Good Service

9-1 海域使用管理情况
Sea Area Use Management

地 区 Region	颁发海域使用权证书（本） Certificates of Right of Sea Area Use Issued (volume)	确权海域面积（公顷） Area of Waters with Established Rights (hm^2)	征收海域使用金（万元） Charge for Sea Area Utilization (10 000 yuan)
全国总计 National Total	**5 011**	**374 148.4**	**850 175.9**
天 津 Tianjin	44	2 061.1	128 129.9
河 北 Hebei	1 250	31 541.9	28 119.1
辽 宁 Liaoning	1 037	103 735.6	115 180.2
上 海 Shanghai	4	489.3	1 507.9
江 苏 Jiangsu	279	38 147.2	60 169.2
浙 江 Zhejiang	214	4 895.6	167 390.1
福 建 Fujian	179	4 583.4	43 204.6
山 东 Shandong	1 464	173 633.8	149 479.2
广 东 Guangdong	133	5 339.9	81 942.6
广 西 Guangxi	359	8 708.0	44 559.7
海 南 Hainan	47	916.8	21 136.7
其 他* others	1	95.9	9 356.7

注：*为沿海省（自治区、直辖市）管理海域以外（指渤海中部海域）。

Note: * Sea areas outside the control of coastal provinces of autonomous regions and municipalities directly under the Central Government. (Referring to the mid-Bohai Sea area)

9-2 海洋倾废管理情况

Management on the Ocean Dumping of Wastes

海 区 Sea Area	签发疏浚物海洋倾倒许可证（份） Permit Issued for Ocean Dumping of Dredged Materials (unit)	新选划倾倒区数（个） Number of Newly Designated Dumping Zones (unit)
合 计 **Total**	**375**	**12**
渤黄海 Bohai and Yellow Sea	25	1
东 海 East China Sea	306	10
南 海 South China Sea	44	1

9-3 海洋行政执法情况
Marine Administrative Enforcement

执法领域 Law Enforcement Area	检查项目（个） Number of Items Subject to Inspection (number)	检查次数（次） Number of Times of Inspection (number)	发现违法行为（起） Illegal Acts Found (case)	作出行政处罚（件） Inflict Administrative Punishments (case)	决定罚款额（万元） Determine Fines (10 000 yuan)	已收缴罚款额（万元） The fine has been Seized (10 000 yuan)
合 计 Total	**48 799**	**145 170**	**1 832**	**1 260**	**167 575.1**	**160 895.2**
海域使用 Sea Area Use	26 504	76 334	975	540	162 194.3	155 732.0
涉外海洋科研 Foreign-related Marine Scientific Research	12	55		4		
海底电缆管道 Submarine Cable Protection	177	1 130	53	284	2.7	2.7
海洋工程建设项目环境保护 Environmental Protection for the Marine Engineering Construction Projects	6 620	32 367	316	124	2 306.1	2 187.5
海洋倾废 Oceanic Dumping of Wastes	1 494	6 960	145	285	692.3	752.3
海洋生态保护 Marine Ecological Conservation	567	5 245	290	23	2 080.1	1 981.1
海岛保护 Island Protection	13 425	23 079	53		299.6	239.6

9-4 沿海地区海滨观测台站分布概况
Distribution of Coastal Observation Stations by Coastal Regions

单位：个 (unit)

地 区 Region	合 计 Total	海洋站 Marine Station	验潮站① Tide Station	气象台站② Meteorological Station	地震台站 Seismic Station	雷达站 Radar Station
合 计 Total	**1 321**	**107**	**252**	**537**	**282**	**143**
天 津 Tianjin	**26**	1		14	7	4
河 北 Hebei	**34**	4		3	21	6
辽 宁 Liaoning	**165**	7	3	115	29	11
上 海 Shanghai	**149**	7	62	56	10	14
江 苏 Jiangsu	**90**	8	26	34	18	4
浙 江 Zhejiang	**140**	20	38	42	21	19
福 建 Fujian	**216**	14	11	123	47	21
山 东 Shandong	**153**	16	22	53	39	23
广 东 Guangdong	**212**	16	82	32	56	26
广 西 Guangxi	**43**	5	5	19	8	6
海 南 Hainan	**93**	9	3	46	26	9

注： ①潮流量观测站43处，潮水位观测站209处。
②中国气象局所属的气象台站仅指沿海气象台站中有海洋气象观测、预报及服务等业务的台站，合计中含有中央气象台1个。

Notes: ①There are 43 tidal current observation stations and 209 tidal level observation stations.
② The meteorological observatories and stations under the China Meteorological Administration only refer to those with marine meteorological observation, forecast and service among the coastal meteorological observatories and the total includes the central meteorological station.

9-5 海洋预报服务概况
Marine Forecast Service

单位：次 (time)

预报项目 Item	数值预报 Numerical Forecast			
	预报服务次数 Frequency	发布次数 Frequency of Release		
		广播电视 Radio and TV	互联网 Internet	纸 质 Paper Media
合 计 Total	**29 395**	**3 111**	**26 459**	**649**
海 浪 Sea Wave	7 300	1 095	6 570	
海 温 Sea Surface Temperature	5 110	730	4 380	
潮 汐 Tide	5 830	730	4 745	365
海 流 Sea Current	4 745		4 745	
海平面 Sea Level				
盐 度 Salinity	730		730	
赤 潮 Red Tide	3		3	
滨海旅游 Coastal Tourism	1 280	185	1 095	
海 冰 Sea Ice	302		302	
绿 潮 Green Seaweed	1 497		1 399	98
溢 油 Oil Spill	22		22	
厄尔尼诺 El Niño				
专 项 Special Item	1 476		1 460	16
其 他 Others	1 100	371	1 008	170

9-5 续表 continued

预报项目 Item	统计预报 Statistical Forecast			
	预报服务次数 Frequency	发布次数 Frequency of Release		
		广播电视 Radio and TV	互联网 Internet	纸质 Paper Media
合　计 Total	**137 904**	**57 066**	**72 248**	**35 583**
海　浪 Sea Wave	42 954	17 648	23 519	9 619
海　温 Sea Surface Temperature	35 690	17 923	19 464	7 427
潮　汐 Tide	31 868	13 804	15 409	7 480
海　流 Sea Current				
海平面 Sea Level				
盐　度 Salinity				
赤　潮 Red Tide	249		107	194
滨海旅游 Coastal Tourism	7 780	5 528	5 086	919
海　冰 Sea Ice	2 066	97	123	1 846
绿　潮 Green Seaweed	58			58
溢　油 Oil Spill	2			2
厄尔尼诺 EL Niño				
专　项 Special Item	15 913	1 095	8 080	7 780
其　他 Others	1 324	971	460	258

注：本表只包含国家海洋局资料。

Note: This table only contains the data from the State Oceanic Administration.

9-6 海洋环境观测情况
Condition of Marine Environmental Observation

项目 Item	台站观测 Station Observation	断面观测 Sectional Observation	浮标观测 Buoy Observation	船舶测报 Ship Measuring and Reporting	其他观测 Other Observation
测站（点）数（个） Number of Stations (unit)	107	119	39	48	191
实际获得数据量（个） Quantity of Data Actually Obtained (unit)	290 486 760	59 735	4 991 614	14 773 855	12 222 695

9-7 海洋调查概况
Marine Survey Statistics

调查名称 Name	站点数（个） Number of Stations (unit)	船舶数（艘） Number of Ships (unit)	项目数（个） Number of Items (unit)	实际获得数据（个） Quantity of Data Actually Obtained (unit)	发布通（公、简）报量（期） Quantity of Circulars (Bulletin, Brief Reports) (unit)
合　计 Total	**2 881**	**704**	**832**	**1 618 492**	**176**
大洋调查 Oceanic Survey	354	7	59	71 560	81
极地调查 Polar Survey	176	3	23	3 181	
专项调查 Special Survey	1 269	234	407	1 048 951	88
其他调查 Other Surveys	1 082	460	343	494 800	7

9-8 涉外海洋科学研究审批情况
Examination and Approval of Foreign-Related Marine Scientific Research

单位：份 (unit)

审批单位 Examination and Approval Authority	研究活动申请审批 Examination and Approval of Application of Research Activities	船只作业计划审批 Examination and Approval of the Ship Operating Plan
国家海洋局 State Oceanic Administration , People's Republic of China	3	3

9-9 海洋档案及利用情况
Marine File and Its Use

指　标 Item	指 标 值 Data
室（馆）存档案 Files Deposited in the Archives	
纸介质档案（卷、册） Paper Media Files (reel,volume)	155 289
电子档案（GB） Electronic Media Files (GB)	21 047
本年接收档案 Files Received this Year	
纸质档案（卷、册） Paper Media Files (reel,volume)	21 640
电子档案（GB） Electronic Media Files (GB)	3 170
本年利用档案 Files Utilized this Year	
纸质档案（件） Paper Media Files (piece)	10 195
电子档案（GB） Electronic Media Files (GB)	2 740

9-10 卫星遥感接收应用情况
Remote-Sensing Receiving and Utilization

指　标 Item	指标值 Data
卫星接收次数（轨） Number of Satellite Receptions (orbit / track)	5 886
全年接收时间（天） Whole Year Receiving Time (day)	5 336
全年实际接收存档数据量（GB） Amount of Data Actually Received and Placed on File in the Whole Year (GB)	41 719.7
累计存档数据量（GB） Total Amount of Data Placed on File (GB)	41 328.6
卫星数据分发 Satellite Data Distribution	
类别用户（个） Classified Users (unit)	
分发数据量（GB） Amount of Data Distributed (GB)	83 855.4

9-11 海洋标准化监督管理情况
Supervision and Management of Marine Standardization

单位：项 (item)

指 标 Item	指标值 Data
标准立项审查 Examination of the Standards for Authorization	211
国家标准 National Standards	31
行业标准 Professional Standards	180
标准审查 Examination of Standards	128
国家标准 National Standards	33
行业标准 Professional Standards	95
标准出版 Standards Publication	23
国家标准 National Standards	14
行业标准 Professional Standards	9
标准实施监督检查（次） Supervision and Examination of Standards Implementation (time)	5

主要统计指标解释

1. 海域使用检查 针对不同类型的用海行为进行的监督检查。

2. 涉外海洋科研项目检查 主要针对国际组织、外国组织和个人为和平目的，单独或者与中华人民共和国的组织合作，使用船舶或者其他运载工具、设施，在中华人民共和国内海、领海以及中华人民共和国管辖的其他海域内进行的对海洋环境和海洋资源等的调查研究活动进行监督检查。

3. 海底电缆管道检查 主要是针对铺设海底电缆管道路由调查、铺设施工和维修改造等的监督检查。

4. 海洋工程建设项目环境保护检查 主要是针对防治海洋工程建设项目对海洋环境的污染损害的监督检查。

5. 海洋倾废检查 主要是针对防治倾倒废弃物对海洋环境的污染损害的监督检查。

6. 海洋生态保护检查 主要是针对红树林、珊瑚礁、滨海湿地、海岛、海湾、入海河口、重要渔业水域等具有典型性、代表性的海洋生态系统，珍稀、濒危海洋生物的天然集中分布区，具有重要经济价值的海洋生物生存区域及有重大科学文化价值的海洋自然历史遗迹和自然景观等海洋自然保护区以及其他需要予以特殊保护的区域的监督检查。

7. 断面监测 按照国家海洋局“断面监测方案”，每年定期利用船舶在沿海设定的断面上进行海洋水文、气象、生物、化学等项目的监测活动。

8. 浮标监测 在海上固定站位获取长期、连续海洋环境观测资料的海上锚定资料浮标。

9. 船舶监测 利用在固定航线上走航的商船或渔船，每日定时所在地的海洋环境状况（主要是水文气象要素），并将监测数据实时发送给有关单位，供海洋环境预报使用。

10. 大洋调查 以大洋科考、研究为目的的远洋调查。

11. 专项调查 为完成国家专项任务进行的海洋调查。

12. 全年接收时间 指全年整个应用系统接收的各类遥感卫星数据时，接收设备工作时间。

13. 全年实际接收存档数据量 指全年整个应用系统接收的各类遥感卫星数据原始数据量。

14. 累计存档数据量 指全年整个应用系统制作的各类遥感产品数据量。

15. 分发数据量 指应用系统提供数据产品用于开展各类应用的数据量。

16. 卫星接收次数 地面观测系统接收卫星数据的数量。

17. 国家标准 针对海洋领域内需要在全国范围内统一的有关技术要求所制定的国家标准。海洋国家标准由国家标准化主管部门统一批准、编号和发布。

18. 行业标准 对没有海洋国家标准而又需要在海洋领域内统一的技术要求所制定的标准。海洋行业标准由国家海洋局统一批准、编号和发布。

Explanatory Notes on Main Statistical Indicators

1. Sea Area Use Supervision and inspection refers to the various types of sea area use conducts.

2. Inspection of Foreign-Related Marine Scientific Research Projects refers to the supervision and inspection of the activities of surveying marine environment and resources conducted by international organizations, foreign organizations and individuals for peaceful purposes, alone or in cooperation with PRC organizations, by using ships or other means of delivery as facilities in the internal seas and territorial waters of the People's Republic of China as well as in the other water under the jurisdiction of the People's Republic of China.

3. Inspection of Submarine Cables and Pipelines is mainly aimed at the survey of the submarine cables and pipelines laying route and supervision and inspection of the laying construction, repair and transformation.

4. Environmental Protection Inspection for the Marine Engineering Construction Projects is mainly directed against preventing and controlling the pollution damage of the marine engineering construction project to the marine environment.

5. Inspection of Oceanic Dumping of Wastes mainly refers to the supervision and inspection aimed at preventing and controlling the pollution damage of wastes dumping to the marine environment.

6. Inspection of Marine Ecological Protection mainly refers to the supervision and inspection of the typical and representative marine ecosystems such as mangroves, coral reef, coastal wetland, sea island, bay, estuaries open to the sea, important fishery waters, etc, the natural centralized distribution zones of rare and endangered marine life, the living areas for the marine life with significant economic values and the marine nature reserves such as the marine natural and historical remains and natural landscapes with important scientific and cultural values as well as other areas needing to be specially protected.

7. Sectional Monitoring According to the "*Sectional Monitoring Plan*" of the State Oceanic Administration, monitoring activities concerning such items as marine hydrology, meteorology, biology and chemistry are carried out regularly every year on the sections set in the coastal area.

8. Buoy Monitoring refers to the monitoring carried out by the offshore mooring data buoys which acquire long-term, continuous marine environmental observations at the fixed stations at sea.

9. Ship Monitoring Monitoring the marine environmental condition in the sea area of fixed times every day by using the merchant or fishing vessels cruising on the fixed navigation line (mainly the hydro meteorological elements) and transmitting the monitored data in real time to the related units for use in the marine environmental forecast.

10. Oceanic Survey refers to the oceanic surveys aimed at the oceanic scientific investigations and research.

11. Special-Subject Survey refers to the oceanic investigation for the purpose of fulfilling the state's special tasks.

12. Whole Year Receiving Time refers to the operating time of receiving equipment in the whole year when the whole application system receives all types of remote sensing satellite data.

13. Amount of Data Actually Received and Placed on File Throughout the year refers to the raw data of remote-sensing satellite data of various kinds received by the whole application system throughout the year.

14. Total Amount of Date Placed on File refers to the date amount of various remote-sensing products made by the whole application systems throughout the year.

15. Amount of Data Distributed refers to the amount of data products provided by the application system for various uses.

16. Number of Satellite Receptions refers to the amount of data received by the ground observation system.

17. National Standards refers to the standards formulated in view of the relevant technical requirements in the marine field that need to be unified throughout the country. The Marine National standards are approved, numbered and issued uniformly by the state department responsible for standardization.

18. Professional Standards refers to the standards formulated for the technical requirements which have no national standards but need to be unified in the marine field. The Marine Professional Standards are approved, numbered and issued by the State Oceanic Administration.

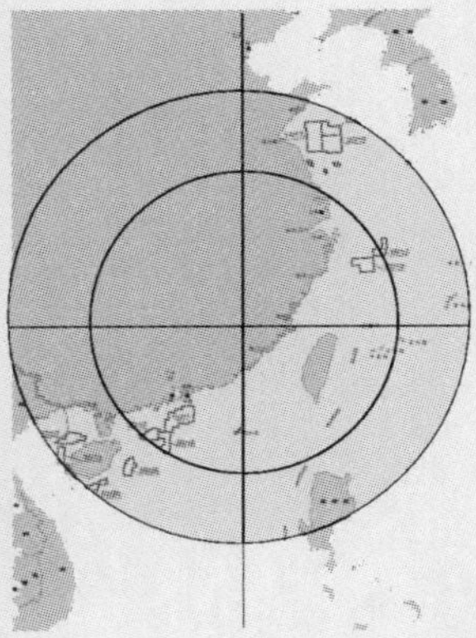

10

全国及沿海社会经济

National and Coastal Socioeconomy

10-1 国内生产总值
Gross Domestic Product

单位：亿元 (100 million yuan)

年 份 Year	国内生产总值 Gross Domestic Product	第一产业 Primary Industry	第二产业 Secondary Industry	第三产业 Tertiary Industry
2001	110 270.4	15 501.2	46 262.0	45 507.2
2002	121 002.0	16 188.6	53 624.4	51 189.0
2003	136 564.6	16 968.3	62 120.8	57 475.6
2004	160 714.4	20 901.8	73 529.8	66 282.8
2005	185 895.8	21 803.5	87 127.3	76 964.9
2006	217 656.6	23 313.0	103 163.5	91 180.1
2007	268 019.4	27 783.0	125 145.4	115 090.9
2008	316 751.7	32 747.0	148 097.9	135 906.9
2009	345 629.2	34 154.0	157 850.1	153 625.1
2010	408 903.0	39 354.6	188 804.9	180 743.4
2011	484 123.5	46 153.3	223 390.3	214 579.9
2012	534 123.0	50 892.7	240 200.4	243 030.0
2013	588 018.8	55 321.7	256 810.0	275 887.0
2014	636 138.7	58 336.1	271 764.5	306 038.2

10-2 国内生产总值增长速度
Growth Rate of Gross Domestic Product

单位：% (%)

年 份 Year	国内生产总值 Gross Domestic Product	第一产业 Primary Industry	第二产业 Secondary Industry	第三产业 Tertiary Industry
2001	8.3	2.6	8.4	10.2
2002	9.1	2.7	9.8	10.5
2003	10.0	2.4	12.7	9.5
2004	10.1	6.1	11.1	10.1
2005	11.3	5.1	12.1	12.3
2006	12.7	4.8	13.4	14.1
2007	14.2	3.5	15.0	16.1
2008	9.6	5.2	9.8	10.5
2009	9.2	4.0	10.1	9.5
2010	10.6	4.3	12.7	9.7
2011	9.5	4.2	10.6	9.5
2012	7.7	4.5	8.2	8.0
2013	7.7	3.8	7.9	8.3
2014	7.3	4.1	7.3	7.8

注：本表按可比价格计算（上年为基期）。

Note:This table is calculated at the comparable price (with the previous year as the base period).

10-3 沿海地区生产总值
Gross Regional Product of Coastal Regions

单位：亿元 (100 million yuan)

地 区 Region	地区生产总值 Gross Regional Product	第一产业 Primary Industry	第二产业 Secondary Industry	第三产业 Tertiary Industry
合 计 Total	**373 069.5**	**24 671.8**	**176 250.8**	**172 146.9**
天 津 Tianjin	15 726.9	199.9	7 731.9	7 795.2
河 北 Hebei	29 421.2	3 447.5	15 012.9	10 960.8
辽 宁 Liaoning	28 626.6	2 285.8	14 384.6	11 956.2
上 海 Shanghai	23 567.7	124.3	8 167.7	15 275.7
江 苏 Jiangsu	65 088.3	3 634.3	30 854.5	30 599.5
浙 江 Zhejiang	40 173.0	1 777.2	19 175.1	19 220.8
福 建 Fujian	24 055.8	2 014.8	12 515.4	9 525.6
山 东 Shandong	59 426.6	4 798.4	28 788.1	25 840.1
广 东 Guangdong	67 809.9	3 166.8	31 419.8	33 223.3
广 西 Guangxi	15 672.9	2 413.4	7 325.0	5 934.5
海 南 Hainan	3 500.7	809.5	876.0	1 815.2

10-4 沿海地区生产总值增长速度
Growth Rate of Gross Regional Product of Coastal Regions

单位：% (%)

地区 Region	2005	2006	2007	2008	2009	2010	2011	2012	2013	2014
天津 Tianjin	14.9	14.7	15.5	16.5	16.5	17.4	16.4	13.8	12.5	10.0
河北 Hebei	13.4	13.4	12.8	10.1	10.0	12.2	11.3	9.6	8.2	6.5
辽宁 Liaoning	12.7	14.2	15.0	13.4	13.1	14.2	12.2	9.5	8.7	5.8
上海 Shanghai	11.4	12.7	15.2	9.7	8.2	10.3	8.2	7.5	7.7	7.0
江苏 Jiangsu	14.5	14.9	14.9	12.7	12.4	12.7	11.0	10.1	9.6	8.7
浙江 Zhejiang	12.8	13.9	14.7	10.1	8.9	11.9	9.0	8.0	8.2	7.6
福建 Fujian	11.6	14.8	15.2	13.0	12.3	13.9	12.3	11.4	11.0	9.9
山东 Shandong	15.0	14.7	14.2	12.0	12.2	12.3	10.9	9.8	9.6	8.7
广东 Guangdong	14.1	14.8	14.9	10.4	9.7	12.4	10.0	8.2	8.5	7.8
广西 Guangxi	13.1	13.6	15.1	12.8	13.9	14.2	12.3	11.3	10.2	8.5
海南 Hainan	10.5	13.2	15.8	10.3	11.7	16.0	12.0	9.1	9.9	8.5

注：本表按可比价格计算（上年为基期）。

Note: This table is calculated at the comparable price (with the previous year as the base period).

10-5 沿海城市生产总值（2013年）
Gross Regional Product of Coastal Cities, 2013

单位：亿元 (100 million yuan)

沿海城市 Coastal City		地区生产总值 Gross Regional Product	第一产业 Primary Industry	第二产业 Secondary Industry	第三产业 Tertiary Industry
合　计	**Total**	**204 074.7**	**10 303.8**	**81 799.3**	**75 712.0**
天　津	**Tianjin**	**14 442.0**			
河　北	**Hebei**	**10 303.0**	**1 036.2**	**5 612.7**	**3 654.1**
唐　山	Tangshan	6 121.2	553.0	3 593.1	1 975.1
秦皇岛	Qinhuangdao	1 168.8	170.1	444.8	553.9
沧　州	Cangzhou	3 013.0	313.1	1 574.8	1 125.1
辽　宁	**Liaoning**	**13 742.3**	**1 153.2**	**7 168.6**	**5 420.6**
大　连	Dalian	7 650.8	477.6	3 892.0	3 281.3
丹　东	Dandong	1 107.3	147.8	547.3	412.1
锦　州	Jinzhou	1 344.9	202.1	656.7	486.1
营　口	Yingkou	1 513.1	109.6	797.7	605.9
盘　锦	Panjin	1 351.1	114.5	911.4	325.2
葫芦岛	Huludao	775.1	101.6	363.5	310.0
上　海	**Shanghai**	**21 818.2**			
江　苏	**Jiangsu**	**10 299.8**	**1 093.8**	**5 066.9**	**4 139.1**
南　通	Nantong	5 038.9	345.4	2 623.5	2 070.0
连云港	Lianyungang	1 785.4	259.2	807.4	718.8
盐　城	Yancheng	3 475.5	489.2	1 636.0	1 350.3
浙　江	**Zhejiang**	**30 675.5**	**1 315.1**	**15 175.9**	**14 184.4**
杭　州	Hangzhou	8 343.5	265.4	3 662.0	4 416.1
宁　波	Ningbo	7 128.9	276.4	3 741.7	3 110.8
温　州	Wenzhou	4 003.9	115.4	2 015.5	1 873.0
嘉　兴	Jiaxing	3 147.7	155.6	1 726.7	1 265.3
绍　兴	Shaoxing	3 967.3	193.3	2 102.9	1 671.1
舟　山	Zhoushan	930.9	95.7	411.5	423.6
台　州	Taizhou	3 153.3	213.3	1 515.6	1 424.5
福　建	**Fujian**	**17 732.3**	**1 283.1**	**9 298.2**	**7 151.0**
福　州	Fuzhou	4 678.5	402.3	2 133.6	2 142.6
厦　门	Xiamen	3 018.2	26.0	1 434.8	1 557.4
莆　田	Putian	1 342.9	114.6	783.5	444.8
泉　州	Quanzhou	5 218.0	171.0	3 227.0	1 819.9
漳　州	Zhangzhou	2 236.0	345.5	1 091.7	798.8
宁　德	Ningde	1 238.7	223.7	627.6	387.5

10-5 续表 continued

沿海城市 Coastal City		地区生产总值 Gross Regional Product	第一产业 Primary Industry	第二产业 Secondary Industry	第三产业 Tertiary Industry
山　东	**Shandong**	**27 497.0**	**1 869.7**	**14 475.3**	**11 151.7**
青　岛	Qingdao	8 006.6	352.4	3 641.1	4 012.8
东　营	Dongying	3 250.2	117.2	2 258.4	874.6
烟　台	Yantai	5 613.9	421.0	3 075.1	2 117.8
潍　坊	Weifang	4 420.7	433.1	2 297.4	1 690.2
威　海	Weihai	2 549.7	203.5	1 312.9	1 033.3
日　照	Rizhao	1 500.2	131.5	784.3	584.4
滨　州	Binzhou	2 155.7	211.0	1 106.1	838.6
广　东	**Guangdong**	**54 273.6**	**2 052.0**	**23 724.2**	**28 497.5**
广　州	Guangzhou	15 420.1	228.9	5 227.4	9 963.9
深　圳	Shenzhen	14 500.2	5.2	6 296.8	8 198.1
珠　海	Zhuhai	1 662.4	43.1	849.1	770.2
汕　头	Shantou	1 565.9	87.2	817.8	660.9
江　门	Jiangmen	2 000.2	158.8	1 013.0	828.3
湛　江	Zhanjiang	2 060.0	421.4	814.3	824.2
茂　名	Maoming	2 160.2	373.2	893.4	893.7
惠　州	Huizhou	2 678.4	136.7	1 550.6	991.1
汕　尾	Shanwei	671.8	108.3	315.7	247.8
阳　江	Yangjiang	1 039.8	192.9	513.8	333.1
东　莞	Dongguan	5 490.0	20.1	2 518.9	2 951.1
中　山	Zhongshan	2 638.9	66.9	1 463.7	1 108.4
潮　州	Chaozhou	780.3	54.9	435.9	289.5
揭　阳	Jieyang	1 605.4	154.4	1 013.8	437.2
广　西	**Guangxi**	**2 013.2**	**393.1**	**986.7**	**634.2**
北　海	Beihai	735.0	142.8	373.7	218.5
防城港	Fangchenggang	525.1	68.5	296.1	160.6
钦　州	Qinzhou	753.1	181.8	316.9	255.1
海　南	**Hainan**	**1 277.8**	**107.6**	**290.8**	**879.4**
海　口	Haikou	904.6	58.1	217.0	629.5
三　亚	Sanya	373.2	49.5	73.8	249.9

注：本表各省数据为合计数。

Note: The data for the provinces are the totals.

10-6 县级单位主要统计指标（2013年）
Main Indicators of Regions at County Level, 2013

单位：万元 (10 000 yuan)

沿海县 Coastal County		第一产业增加值 Value-added of Primary Industry	第二产业增加值 Value-added of Secondary Industry	公共财政收入 Public Revenue of Local Governments	公共财政支出 Public Expenditure of Local Governments
合 计	**Total**	**52 917 555**	**223 879 647**	**31 604 271**	**47 367 409**
河 北	**Hebei**	**3 037 739**	**4 667 296**	**490 672**	**1 163 661**
丰 南	Fengnan				
滦 南	Luannan	781 496	1 076 319	96 588	224 158
乐 亭	Leting	811 549	1 089 303	100 710	226 615
唐 海	Tanghai				
昌 黎	Changli	667 320	740 459	79 093	200 359
抚 宁	Funing	445 282	495 038	66 311	160 062
黄 骅	Huanghua	263 022	1 111 785	122 412	252 873
海 兴	Haixing	69 070	154 392	25 558	99 594
辽 宁	**Liaoning**	**6 813 158**	**26 359 456**	**3 709 243**	**4 972 347**
长 海	Changhai	490 986	82 479	50 078	78 354
瓦房店	Wafangdian	900 022	6 930 966	765 600	906 680
普兰店	Pulandian	955 761	4 747 100	486 042	568 341
庄 河	Zhuanghe	1 168 403	4 876 875	515 924	628 968
东 港	Donggang	712 548	2 433 341	358 511	532 991
凌 海	Linghai	527 002	1 570 661	236 018	307 811
盖 州	Gaizhou	365 546	1 082 536	228 021	408 668
大 洼	Dawa	629 633	2 137 464	393 134	559 338
盘 山	Panshan	463 498	1 027 394	371 564	369 550
绥 中	Suizhong	407 036	715 960	171 610	325 800
兴 城	Xingcheng	192 723	754 680	132 741	285 846

10-6 续表1 continued

沿海县 Coastal County		第一产业增加值 Value-added of Primary Industry	第二产业增加值 Value-added of Secondary Industry	公共财政收入 Public Revenue of Local Governments	公共财政支出 Public Expenditure of Local Governments
江　苏	**Jiangsu**	**7 181 700**	**26 520 300**	**5 141 996**	**7 565 063**
海　安	Hai'an	482 200	2 664 700	466 342	645 808
如　东	Rudong	593 500	2 639 300	406 572	671 073
启　东	Qidong	606 300	3 333 500	585 350	688 660
海　门	Haimen	463 000	4 076 000	611 749	663 692
赣　榆	Ganyu	553 900	1 875 500	349 262	594 652
东　海	Donghai	532 500	1 446 000	327 753	539 656
灌　云	Guanyun	517 600	1 155 600	308 766	499 939
灌　南	Guannan	415 500	1 171 800	305 348	477 700
响　水	Xiangshui	367 500	1 004 700	239 849	382 077
滨　海	Binhai	510 200	1 315 400	290 726	530 104
射　阳	Sheyang	701 400	1 372 900	225 000	450 806
东　台	Dongtai	782 200	2 523 400	524 613	734 736
大　丰	Dafeng	655 900	1 941 500	500 666	686 160
浙　江	**Zhejiang**	**6 302 918**	**53 641 797**	**7 576 599**	**9 692 099**
象　山	Xiangshan	579 680	1 664 999	306 867	545 353
宁　海	Ninghai	393 273	2 084 320	331 354	467 116
余　姚	Yuyao	400 600	4 444 737	596 390	745 919
慈　溪	Cixi	492 980	5 935 659	920 508	983 925
奉　化	Fenghua	283 532	1 340 238	265 662	455 230
洞　头	Dongtou	50 513	254 900	38 344	154 061
平　阳	Pingyang	132 117	1 469 225	203 581	328 165
苍　南	Cangnan	263 752	1 710 530	236 529	453 291
瑞　安	Rui'an	191 141	3 195 265	447 415	506 842
乐　清	Yueqing	196 137	3 929 668	512 937	513 504
海　盐	Haiyan	217 844	1 873 774	246 053	256 893
海　宁	Haining	256 223	3 641 870	524 888	511 850
平　湖	Pinghu	178 301	2 879 611	422 233	430 837

10-6 续表2 continued

沿海县 Coastal County		第一产业增加值 Value-added of Primary Industry	第二产业增加值 Value-added of Secondary Industry	公共财政收入 Public Revenue of Local Governments	公共财政支出 Public Expenditure of Local Governments
绍　兴	Shaoxing	325 525	6 131 030	775 655	686 815
上　虞	Shangyu	417 057	3 442 214	431 075	452 132
岱　山	Daishan	275 379	969 406	105 928	314 575
嵊　泗	Shengsi	192 530	109 764	53 670	215 200
玉　环	Yuhuan	270 100	2 413 631	275 990	371 482
三　门	Sanmen	225 255	576 299	118 035	226 580
温　岭	Wenling	581 941	3 466 005	438 479	607 223
临　海	Linhai	379 038	2 108 652	325 006	465 106
福　建	**Fujian**	**7 142 357**	**41 506 929**	**5 216 353**	**7 417 841**
连　江	Lianjiang	1 053 563	1 165 002	378 800	402 642
罗　源	Luoyuan	270 563	1 107 993	122 628	288 585
平　潭	Pingtan	335 467	482 900	133 983	443 679
福　清	Fuqing	843 522	3 415 739	431 953	1 048 673
长　乐	Changle	390 770	3 257 469	283 511	351 839
仙　游	Xianyou	265 006	1 184 099	215 189	336 841
惠　安	Hui'an	275 114	4 082 457	282 218	417 506
石　狮	Shishi	194 762	3 404 648	359 008	417 520
晋　江	Jinjiang	182 779	9 197 616	1 002 874	990 382
南　安	Nan'an	226 336	4 485 848	398 160	504 907
云　霄	Yunxiao	231 935	563 933	58 100	159 645
漳　浦	Zhangpu	558 249	1 093 256	252 859	343 229
诏　安	Zhao'an	342 642	647 195	73 748	189 035
东　山	Dongshan	271 906	599 710	154 146	200 818
龙　海	Longhai	544 919	2 933 764	616 412	501 227
霞　浦	Xiapu	441 303	492 755	74 270	205 370
福　安	Fu'an	378 046	1 903 663	206 511	331 136
福　鼎	Fuding	335 475	1 488 882	171 983	284 807

10-6 续表3 continued

沿海县 Coastal County		第一产业 增加值 Value-added of Primary Industry	第二产业 增加值 Value-added of Secondary Industry	公共财政收入 Public Revenue of Local Governments	公共财政支出 Public Expenditure of Local Governments
山 东	**Shandong**	**8 353 191**	**52 126 074**	**6 419 978**	**8 142 342**
胶 州	Jiaozhou	492 458	4 557 000	566 377	680 827
即 墨	Jimo	589 097	4 542 500	642 985	781 610
胶 南	Jiaonan				
垦 利	Kenli	178 643	2 164 605	181 397	245 678
利 津	Lijin	232 988	1 224 246	101 339	201 925
广 饶	Guangrao	398 643	4 751 944	350 016	442 521
长 岛	Changdao	335 626	57 298	10 518	75 743
龙 口	Longkou	337 059	5 663 241	715 686	715 820
莱 阳	Laiyang	404 392	1 557 696	110 579	253 267
莱 州	Laizhou	644 215	3 494 725	464 507	517 747
蓬 莱	Penglai	282 241	2 493 975	255 449	330 872
招 远	Zhaoyuan	349 816	3 380 531	415 088	459 558
海 阳	Haiyang	572 428	1 329 588	221 236	276 794
寿 光	Shouguang	842 136	3 412 672	706 900	783 814
昌 邑	Changyi	319 849	1 853 866	233 957	283 026
文 登	Wendeng	555 468	3 215 865	414 877	512 307
荣 成	Rongcheng	767 592	4 442 985	522 399	784 778
乳 山	Rushan	348 224	2 118 738	244 768	329 261
无 棣	Wudi	346 307	1 259 224	160 850	273 336
沾 化	Zhanhua	356 009	605 375	101 050	193 458
广 东	**Guangdong**	**7 924 295**	**14 422 681**	**1 515 134**	**4 387 489**
南 澳	Nan'ao	37 367	57 266	16 358	64 465
台 山	Taishan	508 349	1 780 648	203 256	355 062
恩 平	Enping	190 020	446 063	84 286	183 463
遂 溪	Suixi	898 520	717 145	58 193	249 188
徐 闻	Xuwen	648 721	164 189	38 977	216 957

10-6 续表4 continued

沿海县 Coastal County		第一产业增加值 Value-added of Primary Industry	第二产业增加值 Value-added of Secondary Industry	公共财政收入 Public Revenue of Local Governments	公共财政支出 Public Expenditure of Local Governments
廉　江	Lianjiang	873 909	1 266 056	81 608	404 300
雷　州	Leizhou	869 098	308 843	53 393	340 839
吴　川	Wuchuan	254 291	769 251	53 326	228 014
电　白	Dianbai	760 183	1 209 807	125 918	365 717
惠　东	Huidong	386 684	1 861 814	232 642	414 017
海　丰	Haifeng	333 160	1 000 341	132 107	227 856
陆　丰	Lufeng	451 394	923 367	171 689	403 985
阳　西	Yangxi	500 722	533 262	51 388	193 103
阳　东	Yangdong	398 904	1 323 505	101 161	199 344
饶　平	Raoping	343 593	857 676	57 095	260 986
揭　东	Jiedong				
惠　来	Huilai	469 380	1 203 448	53 737	280 193
广　西	**Guangxi**	**830 592**	**741 779**	**155 018**	**469 467**
合　浦	Hepu	704 828	432 960	58 752	287 012
东　兴	Dongxing	125 764	308 819	96 266	182 455
海　南	**Hainan**	**5 331 605**	**3 893 335**	**1 379 278**	**3 557 100**
琼　海	Qionghai	607 107	276 505	177 667	386 510
儋　州	Danzhou	962 137	284 007	115 558	455 763
文　昌	Wenchang	644 854	440 565	185 439	344 839
万　宁	Wanning	414 120	383 144	123 642	350 906
东　方	Dongfang	316 518	607 036	103 398	341 100
澄　迈	Chengmai	556 191	963 818	185 193	406 297
临　高	Lingao	811 628	109 443	54 883	286 202
昌　江	Changjiang	203 552	531 921	101 798	244 288
乐　东	Ledong	478 195	84 225	97 727	326 765
陵　水	Lingshui	337 303	212 671	233 973	414 430

注：本表各省数据为合计数。

Note: The data for the provinces are the totals.

10-7 沿海地区财政收支
Financial Revenue and Expenditure by Coastal Regions

单位：亿元 (100 million yuan)

地 区 Region	地方一般公共预算收入 General Public Budget Revenue	地方一般公共预算支出 General Public Budget Expenditure
合 计 Total	**41 402.2**	**55 414.1**
天 津 Tianjin	2 390.4	2 884.7
河 北 Hebei	2 446.6	4 677.3
辽 宁 Liaoning	3 192.8	5 080.5
上 海 Shanghai	4 585.6	4 923.4
江 苏 Jiangsu	7 233.1	8 472.5
浙 江 Zhejiang	4 122.0	5 159.6
福 建 Fujian	2 362.2	3 306.7
山 东 Shandong	5 026.8	7 177.3
广 东 Guangdong	8 065.1	9 152.6
广 西 Guangxi	1 422.3	3 479.8
海 南 Hainan	555.3	1 099.7

10-8 沿海地区教育基本情况
Basic Conditions of Education by Coastal Regions

地　区 Region	高等学校数 （所） Institutions of Higher Education (unit)	本、专科在校学生数 （人） Number of Students at School in Undergraduate or Specialized Courses (person)	本、专科毕（结）业生数 （人） Number of Students Graduated in Undergraduate or Specialized Courses (person)
全国总计 National Total	**2 529**	**25 476 999**	**6 593 671**
天　津 Tianjin	55	505 795	123 505
河　北 Hebei	118	1 164 341	344 518
辽　宁 Liaoning	116	998 281	247 510
上　海 Shanghai	68	506 644	132 411
江　苏 Jiangsu	159	1 698 636	478 713
浙　江 Zhejiang	104	978 216	253 708
福　建 Fujian	88	748 480	190 144
山　东 Shandong	141	1 796 665	464 076
广　东 Guangdong	141	1 794 188	440 952
广　西 Guangxi	70	701 913	174 050
海　南 Hainan	17	180 565	44 792

10-9 沿海地区卫生基本情况
Basic Conditions of Public Health by Coastal Regions

地 区 Region	卫生机构数 （个） Health Institutions (unit)	医疗机构床位数 （万张） Total Beds (10 000 beds)	卫生机构人员 （人） Number of Employed Personnel in Health Institutions (person)
全国总计 National Total	**981 432**	**660.12**	**10 234 213**
天 津 Tianjin	4 990	6.09	111 672
河 北 Hebei	78 895	32.29	512 877
辽 宁 Liaoning	35 441	25.55	339 187
上 海 Shanghai	4 984	11.75	201 735
江 苏 Jiangsu	31 995	39.23	589 559
浙 江 Zhejiang	30 358	24.58	455 809
福 建 Fujian	28 030	16.48	273 602
山 东 Shandong	77 012	50.06	838 474
广 东 Guangdong	48 085	40.58	732 573
广 西 Guangxi	34 667	20.16	357 622
海 南 Hainan	5 075	3.45	66 556

10-10 分地区电力消费量
Electricity Consumption by Region

单位：亿千瓦·时 (100 million kW·h)

地 区 Region	2012	2013	2014
天 津 Tianjin	722.5	774.5	794.4
河 北 Hebei	3 078.0	3 251.2	3 314.1
辽 宁 Liaoning	1 899.9	2 008.5	2 038.7
上 海 Shanghai	1 353.5	1 410.6	1 369.0
江 苏 Jiangsu	4 580.9	4 956.6	5 012.5
浙 江 Zhejiang	3 210.6	3 453.1	3 506.4
福 建 Fujian	1 579.5	1 700.7	1 855.8
山 东 Shandong	3 794.6	4 083.1	4 223.5
广 东 Guangdong	4 619.4	4 830.1	5 235.2
广 西 Guangxi	1 153.9	1 237.7	1 308.0
海 南 Hainan	210.9	232.0	251.9

10-11 沿海地区用水情况
Water Use of Coastal Regions

地 区 Region	全年供水总量 （万立方米） Total Annual Volume of Water Supply (10 000 m^3)	人均用水量 （立方米/人） Per Capita Water Use (m^3/person)
全国总计 **National Total**	**6 094.9**	**446.7**
天 津 Tianjin	24.1	161.2
河 北 Hebei	192.8	262.0
辽 宁 Liaoning	141.8	322.9
上 海 Shanghai	105.9	437.6
江 苏 Jiangsu	591.3	743.8
浙 江 Zhejiang	192.9	
福 建 Fujian	205.6	542.6
山 东 Shandong	214.5	219.8
广 东 Guangdong	442.5	414.2
广 西 Guangxi	307.6	649.4
海 南 Hainan	45.0	500.7

10-12 沿海地区全社会固定资产投资
Total Investment in Fixed Assets by the Whole Society of Coastal Regions

单位：亿元　　(100 million yuan)

地 区 Region	全社会固定资产投资 Total Investment in Fixed Assets in the Whole Country	#农、林、牧、渔业 Agriculture, Forestry, Animal Husbandry and Fishery	#交通运输、仓储和邮政业 Transport, Storage and Post
全国总计 National Total	**512 020.7**	**16 573.8**	**43 215.7**
天 津 Tianjin	10 518.2	214.3	750.6
河 北 Hebei	26 671.9	1 204.3	2 046.5
辽 宁 Liaoning	24 730.8	662.2	1 808.5
上 海 Shanghai	6 016.4	11.9	459.2
江 苏 Jiangsu	41 938.6	323.3	2 169.7
浙 江 Zhejiang	24 262.8	305.0	1 736.6
福 建 Fujian	18 177.9	472.8	1 783.7
山 东 Shandong	42 495.5	1 166.3	2 196.1
广 东 Guangdong	26 293.9	366.6	2 589.9
广 西 Guangxi	13 843.2	630.9	1 301.6
海 南 Hainan	3 112.2	47.4	351.2

10-13 沿海地区按经营单位所在地分货物进出口总额
Total Value of Imports and Exports by Location of Importers/Exporters of Coastal Regions

单位：万美元 (10 000 USD)

地 区 Region	进出口 Total	出口 Exports	进口 Imports
全国总计 National Total	**430 152 735**	**234 229 270**	**195 923 465**
天 津 Tianjin	13 388 608	5 259 066	8 129 542
河 北 Hebei	5 987 736	3 571 020	2 416 716
辽 宁 Liaoning	11 399 826	5 874 518	5 525 307
上 海 Shanghai	46 639 984	21 013 386	25 626 598
江 苏 Jiangsu	56 355 308	34 183 250	22 172 058
浙 江 Zhejiang	35 503 977	27 332 705	8 171 272
福 建 Fujian	17 740 784	11 345 229	6 395 555
山 东 Shandong	27 692 940	14 470 865	13 222 074
广 东 Guangdong	107 658 447	64 608 701	43 049 746
广 西 Guangxi	4 054 885	2 432 743	1 622 143
海 南 Hainan	1 586 266	441 674	1 144 592

10-14 沿海地区城镇居民平均每人全年家庭收入和消费性支出
Per Capita Annual Income and Consumption Expenditure of Urban Households by Coastal Regions

单位：元 (yuan)

地区 Region	可支配收入 Disposable Income	工资性收入 Income from Wages and Salaries	消费支出 Expenses on Consumption
全国平均水平 National Average	**28 843.9**	**17 936.8**	**19 968.1**
天津 Tianjin	31 506.0	18 796.9	24 289.6
河北 Hebei	24 141.3	15 275.9	16 203.8
辽宁 Liaoning	29 081.7	16 239.6	20 519.6
上海 Shanghai	48 841.4	30 212.1	35 182.4
江苏 Jiangsu	34 346.3	20 720.1	23 476.3
浙江 Zhejiang	40 392.7	23 317.3	27 241.7
福建 Fujian	30 722.4	19 197.2	22 204.1
山东 Shandong	29 221.9	18 866.2	18 322.6
广东 Guangdong	32 148.1	24 315.6	23 611.7
广西 Guangxi	24 669.0	13 892.7	15 045.4
海南 Hainan	24 486.5	15 654.2	17 513.8

10-15 沿海地区农村居民家庭人均纯收入和生活消费支出
Per Capita Annual Net Income and Consumption Expenditure of Rural Households by Coastal Regions

单位：元 (yuan)

地　区 Region	可支配收入 Disposable Income	消费支出 Consumption Expenditure
全国平均水平 **National Average**	**10 488.9**	**8 382.6**
天　津 Tianjin	17 014.2	13 738.6
河　北 Hebei	10 186.1	8 248.0
辽　宁 Liaoning	11 191.5	7 800.7
上　海 Shanghai	21 191.6	14 820.1
江　苏 Jiangsu	14 958.4	11 820.3
浙　江 Zhejiang	19 373.3	14 497.8
福　建 Fujian	12 650.2	11 055.9
山　东 Shandong	11 882.3	7 962.2
广　东 Guangdong	12 245.6	10 043.2
广　西 Guangxi	8 683.2	6 675.1
海　南 Hainan	9 912.6	7 029.0

10-16 沿海地区年末人口数
Year-end Population by Coastal Regions

单位：万人 (10 000 persons)

地 区 Region	2007	2008	2009	2010	2011	2012	2013	2014
全国总计 National Total	**132 129**	**132 802**	**133 450**	**134 091**	**134 735**	**135 404**	**136 072**	**136 782**
天 津 Tianjin	1 115	1 176	1 228	1 299	1 355	1 413	1 472	1 517
河 北 Hebei	6 943	6 989	7 034	7 194	7 241	7 288	7 333	7 384
辽 宁 Liaoning	4 298	4 315	4 341	4 375	4 383	4 389	4 390	4 391
上 海 Shanghai	2 064	2 141	2 210	2 303	2 347	2 380	2 415	2 426
江 苏 Jiangsu	7 723	7 762	7 810	7 869	7 899	7 920	7 939	7 960
浙 江 Zhejiang	5 155	5 212	5 276	5 447	5 463	5 477	5 498	5 508
福 建 Fujian	3 612	3 639	3 666	3 693	3 720	3 748	3 774	3 806
山 东 Shandong	9 367	9 417	9 470	9 588	9 637	9 685	9 733	9 789
广 东 Guangdong	9 660	9 893	10 130	10 441	10 505	10 594	10 644	10 724
广 西 Guangxi	4 768	4 816	4 856	4 610	4 645	4 682	4 719	4 754
海 南 Hainan	845	854	864	869	877	887	895	903

注：2010年数据为当年人口普查数据推算数；其余年份数据为年度人口抽样调查推算数据。

Note: Data of 2010 are the census year estimates; the rest are the estimates from the annual national sample survey on population.

10-17 沿海城市人口情况（2013年）
Population of Coastal Cities, 2013

单位：万人　　　　(10 000 persons)

沿海城市	Coastal City	常住人口 Permanent Population	年末总人口 Total Population by the End of the Year
合　计	**Total**	**25 229.7**	**22 261.5**
天　津	**Tianjin**		
河　北	**Hebei**	**1 806.3**	**1 770.9**
唐　山	Tangshan	770.8	738.7
秦皇岛	Qinhuangdao	304.5	290.7
沧　州	Cangzhou	731.0	741.5
辽　宁	**Liaoning**	**1 892.4**	**1 778.3**
大　连	Dalian	694.3	591.4
丹　东	Dandong	242.6	239.6
锦　州	Jinzhou	308.7	305.9
营　口	Yingkou	244.4	232.5
盘　锦	Panjin	143.8	129.0
葫芦岛	Huludao	258.6	279.9
上　海	**Shanghai**		
江　苏	**Jiangsu**	**1 894.6**	**2 110.5**
南　通	Nantong	729.8	766.5
连云港	Lianyungang	442.8	520.2
盐　城	Yancheng	722.0	823.8

10-17 续表1 continued

沿海城市 Coastal City		常住人口 Permanent Population	年末总人口 Total Population by the End of the Year
浙　江	**Zhejiang**	**4 239.0**	**3 572.8**
杭　州	Hangzhou	884.4	706.6
宁　波	Ningbo	766.3	580.1
温　州	Wenzhou	919.7	807.2
嘉　兴	Jiaxing	455.7	345.9
绍　兴	Shaoxing	494.9	441.7
舟　山	Zhoushan	114.2	97.3
台　州	Taizhou	603.8	594.0
福　建	**Fujian**	**3 003.0**	**2 694.9**
福　州	Fuzhou	734.0	623.7
厦　门	Xiamen	373.0	196.8
莆　田	Putian	283.0	334.2
泉　州	Quanzhou	836.0	703.5
漳　州	Zhangzhou	493.0	489.5
宁　德	Ningde	284.0	347.2
山　东	**Shandong**	**3 672.6**	**3 420.3**
青　岛	Qingdao	896.4	773.7
东　营	Dongying	208.5	187.0
烟　台	Yantai	698.9	651.2
潍　坊	Weifang	922.5	882.9
威　海	Weihai	280.6	253.8
日　照	Rizhao	285.1	290.1
滨　州	Binzhou	380.6	381.6

10-17 续表2 continued

沿海城市 Coastal City		常住人口 Permanent Population	年末总人口 Total Population by the End of the Year
广　东	**Guangdong**	**7 866.7**	**6 034.0**
广　州	Guangzhou	1 292.7	832.3
深　圳	Shenzhen	1 062.9	324.3
珠　海	Zhuhai	159.0	108.6
汕　头	Shantou	547.9	540.0
江　门	Jiangmen	449.8	393.0
湛　江	Zhanjiang	716.7	804.2
茂　名	Maoming	601.3	757.7
惠　州	Huizhou	470.0	343.4
汕　尾	Shanwei	298.6	352.5
阳　江	Yangjiang	248.0	285.1
东　莞	Dongguan	831.7	188.9
中　山	Zhongshan	317.4	154.1
潮　州	Chaozhou	271.2	267.2
揭　阳	Jieyang	599.5	682.7
广　西	**Guangxi**	**564.8**	**658.9**
北　海	Beihai	159.0	169.4
防城港	Fangchenggang	89.9	93.0
钦　州	Qinzhou	315.9	396.5
海　南	**Hainan**	**290.3**	**220.9**
海　口	Haikou	217.1	163.2
三　亚	Sanya	73.2	57.7

注：本表各省数据为合计数。

Note: The data for the provinces are the totals.

10-18 沿海县人口情况（2013年）
Population of Coastal Counties, 2013

单位：万人　　(10 000 persons)

沿海县	Coastal County	户籍总人口 Total Population of Household Registration
合　计	**Total**	**8 483**
河　北	**Hebei**	**280**
滦　南	Luannan	57
乐　亭	Leting	49
唐　海	Tanghai	
昌　黎	Changli	56
抚　宁	Funing	48
黄　骅	Huanghua	47
海　兴	Haixing	23
辽　宁	**Liaoning**	**659**
长　海	Changhai	7
瓦房店	Wafangdian	100
普兰店	Pulandian	93
庄　河	Zhuanghe	90
东　港	Donggang	61
凌　海	Linghai	52
盖　州	Gaizhou	70
大　洼	Dawa	39
盘　山	Panshan	28
绥　中	Suizhong	65
兴　城	Xingcheng	54

10-18 续表1 continued

沿海县 Coastal County		户籍总人口 Total Population of Household Registration
江　苏	**Jiangsu**	**1 298**
海　安	Hai'an	94
如　东	Rudong	104
启　东	Qidong	112
海　门	Haimen	100
赣　榆	Ganyu	118
东　海	Donghai	120
灌　云	Guanyun	104
灌　南	Guannan	80
响　水	Xiangshui	62
滨　海	Binhai	120
射　阳	Sheyang	97
东　台	Dongtai	114
大　丰	Dafeng	73
浙　江	**Zhejiang**	**1 484**
象　山	Xiangshan	54
宁　海	Ninghai	62
余　姚	Yuyao	84
慈　溪	Cixi	104
奉　化	Fenghua	48
洞　头	Dongtou	13
平　阳	Pingyang	88
苍　南	Cangnan	132
瑞　安	Rui'an	122
乐　清	Yueqing	128
海　盐	Haiyan	38
海　宁	Haining	67
平　湖	Pinghu	49

10-18 续表2 continued

沿海县 Coastal County		户籍总人口 Total Population of Household Registration
绍 兴	Shaoxing	64
上 虞	Shangyu	78
岱 山	Daishan	19
嵊 泗	Shengsi	8
玉 环	Yuhuan	43
三 门	Sanmen	44
温 岭	Wenling	121
临 海	Linhai	118
福 建	**Fujian**	**1 316**
连 江	Lianjiang	65
罗 源	Luoyuan	26
平 潭	Pingtan	42
福 清	Fuqing	132
长 乐	Changle	70
仙 游	Xianyou	111
惠 安	Hui'an	98
石 狮	Shishi	32
晋 江	Jinjiang	109
南 安	Nan'an	154
云 霄	Yunxiao	44
漳 浦	Zhangpu	87
诏 安	Zhao'an	62
东 山	Dongshan	21
龙 海	Longhai	84
霞 浦	Xiapu	54
福 安	Fu'an	66
福 鼎	Fuding	59

10-18 续表3 continued

沿海县 Coastal County		户籍总人口 Total Population of Household Registration
山 东	**Shandong**	**1 142**
胶 州	Jiaozhou	82
即 墨	Jimo	114
胶 南	Jiaonan	
垦 利	Kenli	22
利 津	Lijin	30
广 饶	Guangrao	50
长 岛	Changdao	4
龙 口	Longkou	64
莱 阳	Laiyang	87
莱 州	Laizhou	85
蓬 莱	Penglai	45
招 远	Zhaoyuan	57
海 阳	Haiyang	66
寿 光	Shouguang	106
昌 邑	Changyi	58
文 登	Wendeng	64
荣 成	Rongcheng	67
乳 山	Rushan	56
无 棣	Wudi	46
沾 化	Zhanhua	39
广 东	**Guangdong**	**1 639**
南 澳	Nan'ao	8
台 山	Taishan	98
恩 平	Enping	50
遂 溪	Suixi	107
徐 闻	Xuwen	72

10-18 续表4 continued

沿海县 Coastal County		户籍总人口 Total Population of Household Registration
廉　江	Lianjiang	175
雷　州	Leizhou	172
吴　川	Wuchuan	116
电　白	Dianbai	144
惠　东	Huidong	86
海　丰	Haifeng	83
陆　丰	Lufeng	185
阳　西	Yangxi	52
阳　东	Yangdong	49
饶　平	Raoping	105
揭　东	Jiedong	
惠　来	Huilai	137
广　西	**Guangxi**	**120**
东　兴	Dongxing	14
合　浦	Hepu	106
海　南	**Hainan**	**545**
琼　海	Qionghai	51
儋　州	Danzhou	99
文　昌	Wenchang	60
万　宁	Wanning	63
东　方	Dongfang	46
澄　迈	Chengmai	57
临　高	Lingao	51
昌　江	Changjiang	26
乐　东	Ledong	55
陵　水	Lingshui	37

注：本表各省数据为合计数。

Note: The data for the provinces are the totals.

10-19 沿海地区城镇单位就业人员情况
Employed Persons in Urban Units by Coastal Regions

单位：万人 (10 000 persons)

地 区 Region	2012	2013	2014
全国总计 National Total	**15 236.4**	**18 108.4**	**18 277.8**
天 津 Tianjin	289.1	302.4	295.5
河 北 Hebei	619.9	653.4	656.2
辽 宁 Liaoning	598.7	689.1	665.2
上 海 Shanghai	555.7	618.8	648.9
江 苏 Jiangsu	830.9	1 503.3	1 602.4
浙 江 Zhejiang	1 070.1	1 071.6	1 102.7
福 建 Fujian	637.9	644.0	654.6
山 东 Shandong	1 110.2	1 290.6	1 266.3
广 东 Guangdong	1 304.0	1 967.0	1 973.3
广 西 Guangxi	358.0	403.0	401.5
海 南 Hainan	90.1	98.8	101.5

10-20 沿海城市就业人员情况（2012年）
Number of Employed Persons by Coastal Cities, 2012

单位：万人 (10 000 persons)

沿海城市 Coastal City	就业人员 Number of Employed Persons	城镇就业人员 Urban Employed Persons
合　计　Total	**15 745.7**	**4 573.5**
天　津　Tianjin		**73.4**
河　北　Hebei	**1 045.0**	**183.1**
唐　山　Tangshan	451.9	96.5
秦皇岛　Qinhuangdao	170.1	34.2
沧　州　Cangzhou	423.0	52.4
辽　宁　Liaoning	**1 183.7**	**301.4**
大　连　Dalian	472.8	131.4
丹　东　Dandong	132.9	31.8
锦　州　Jinzhou	170.6	32.1
营　口　Yingkou	154.0	29.1
盘　锦　Panjin	105.5	49.8
葫芦岛　Huludao	147.9	27.2
上　海　Shanghai		
江　苏　Jiangsu	**1 163.8**	**321.0**
南　通　Nantong	467.2	184.6
连云港　Lianyungang	250.2	47.4
盐　城　Yancheng	446.4	89.0
浙　江　Zhejiang	**2 869.9**	**893.1**
杭　州　Hangzhou	650.5	282.6
宁　波　Ningbo	503.4	171.4
温　州　Wenzhou	574.0	103.0
嘉　兴　Jiaxing	327.7	79.8
绍　兴　Shaoxing	344.4	135.5
舟　山　Zhoushan	72.7	18.5
台　州　Taizhou	397.2	102.3
福　建　Fujian	**2 018.8**	**561.7**
福　州　Fuzhou	479.1	142.8
厦　门　Xiamen	257.2	130.3

10-20 续表 continued

沿海城市 Coastal City		就业人员 Number of Employed Persons	城镇就业人员 Urban Employed Persons
莆　田	Putian	208.8	46.6
泉　州	Quanzhou	578.9	163.7
漳　州	Zhangzhou	299.7	50.6
宁　德	Ningde	195.1	27.7
山　东	**Shandong**	**2 467.5**	**537.0**
青　岛	Qingdao	571.4	147.1
东　营	Dongying	148.2	48.8
烟　台	Yantai	493.3	107.9
潍　坊	Weifang	563.1	92.7
威　海	Weihai	198.4	56.0
日　照	Rizhao	207.7	32.1
滨　州	Binzhou	285.4	52.4
广　东	**Guangdong**	**4 640.4**	**1 598.1**
广　州	Guangzhou	759.9	324.6
深　圳	Shenzhen	899.2	457.4
珠　海	Zhuhai	106.3	74.6
汕　头	Shantou	239.7	55.5
江　门	Jiangmen	244.3	59.5
湛　江	Zhanjiang	336.4	50.3
茂　名	Maoming	280.5	44.8
惠　州	Huizhou	277.3	86.1
汕　尾	Shanwei	119.7	24.9
阳　江	Yangjiang	129.0	24.5
东　莞	Dongguan	633.2	244.9
中　山	Zhongshan	210.3	90.8
潮　州	Chaozhou	131.0	21.1
揭　阳	Jieyang	273.6	39.1
广　西	**Guangxi**	**166.1**	**44.2**
北　海	Beihai	70.1	13.8
防城港	Fangchenggang	61.4	11.0
钦　州	Qinzhou	34.6	19.4
海　南	**Hainan**	**190.5**	**60.5**
海　口	Haikou	153.0	49.4
三　亚	Sanya	37.5	11.1

注：本表各省数据为合计数。
Note: The data for the provinces are the totals.

主要统计指标解释

1. 国内(或地区)生产总值　指一个国家（或地区）所有常驻单位在一定时期内生产活动的最终成果。国内生产总值有三种表现形态，即价值形态、收入形态和产品形态。从价值形态看，它是所有常驻单位在一定时期内生产的全部货物和服务价值超过同期中间投入的全部非固定资产货物和服务价值的差额，即所有常驻单位的增加值之和；从收入形态看，它是所有常驻单位在一定时期内创造并分配给常驻单位和非常驻单位的初次收入分配之和；从产品形态看，它是所有常驻单位在一定时期内最终使用的货物和服务价值与货物和服务净出口价值之和。在实际核算中，国内生产总值有三种计算方法，即生产法、收入法和支出法。三种方法分别从不同的方面反映国内生产总值及其构成。

2. 三次产业　是根据社会生产活动历史发展的顺序对产业结构的划分，产品直接取自自然界的部门称为第一产业，对初级产品进行再加工的部门称为第二产业，为生产和消费提供各种服务的部门称为第三产业。它是世界上较为通用的产业结构分类，但各国的划分不尽一致。我国的三次产业划分如下。

第一产业：是指农、林、牧、渔业。

第二产业：是指采矿业，制造业，电力、燃气及水的生产和供应业，建筑业。

第三产业：是指除第一、第二产业以外的其他行业。第三产业包括：交通运输、仓储和邮政业，信息传输、计算机服务和软件业，批发和零售业，住宿和餐饮业，金融业，房地产业，租赁和商务服务业，科学研究、技术服务和地质勘查业，水利、环境和公共设施管理业，居民服务和其他服务业，教育，卫生、社会保障和社会福利业，文化、体育和娱乐业，公共管理和社会组织，国际组织。

3. 增加值　是指各行各业生产经营和劳务活动的最终成果，采用生产法和收入法两种方法计算。解生产法　是从货物和服务活动在生产过程中形成的总产品入手，剔除生产过程中投入的中间产品价值，得到新增价值的方法。

收入法　又称分配法。按收入法计算国内生产总值是从生产过程创造的收入的角度对常驻单位的生产活动成果进行核算；按照此法计算，增加值由劳动者报酬、固定资产折旧、生产税净额和营业盈余四个部分组成。解

4. 年末总人口　是指每年 12 月 31 日 24 时一定地区范围内的有生命的个人的人口总和。

5. 财政收入　指国家财政参与社会产品分配所取得的收入，是实现国家职能的财力保证。财政收入所包括的内容几经变化，目前主要包括：

（1）各项税收　包括增值税、营业税、消费税、土地增值税、城市维护建设税、资源税、城市土地使用税、企业所得税、个人所得税、关税、证券交易印花税、车辆购置税、农牧业税和耕地占用税等；

（2）专项收入　包括排污费收入、城市水资源费收入、矿产资源补偿费收入、教育费附加收入等；

（3）其他收入　包括利息收入、基本建设贷款归还收入、基本建设收入、捐赠收入等；

（4）国有企业亏损补贴　此项为负收入，冲减财政收入。主要包括对工业企业、商业企业、粮食企业的补贴。

6. 财政支出　国家财政将筹集起来的资金进行分配使用，以满足经济建设和各项事业的需要。

7. 基本建设支出　指按国家有关规定，属于基本建设范围内的基本建设有偿使用、拨款、资本金支出以及经国家批准对专项和政策性基建投资贷款，在部门的基建投资额中统筹支付的贴息支出。

8. 普通高等学校　指按照国家规定的设置标准和审批程序批准举办的，通过全国普通高等学校统一招生考试，招收高中毕业生为主要培养对象，实施高等教育的全日制大学、独立设置的学院和高等专科学校、高等职业学校和其他机构。

大学、独立设置的学院主要实施本科层次以上教育，高等专科学校、高等职业学校实施专科层次教育，其他机构是承担国家普通招生计划任务不计校数的机构。包括普通高等学校分校和批准筹建的普通高等学校等。

9. 卫生机构　包括医疗机构、疾病预防控制中心(防疫站)、采供血机构、卫生监督及监测(检验)机构、医学科研和在职培训机构、健康教育所等。

10. 医疗机构　包括医院、社区卫生服务中心(站)、疗养院、卫生院、门诊部、诊所(卫生所、医务室)、妇幼保健院(所、站)、专科疾病防治院(所、站)、急救中心(站)和临床检验中心。医疗机构分为非营利性医疗机构和营利性医疗机构。

11. 用水总量　指分配给各类用户的包括输水损失在内的毛用水量之和，不包括海水直接利用量。

12. 全社会固定资产投资　是以货币形式表现的在一定时期内全社会建造和购置固定资产的工作量以及与此有关的费用的总称。该指标是反映固定资产投资规模、结构和发展速度的综合性指标，又是观察工程进度和考核投资效果的重要依据。全社会固定资产投资按登记注册类型可分为国有、集体、个体、联营、股份制、外商、港澳台商、其他等。

13. 进出口总额　指实际进出我国国境的货物总金额。包括对外贸易实际进出口货物，来料加工装配进出口货物，国家间、联合国及国际组织无偿援助物资和赠送品，华侨、港澳台同胞和外籍华人捐赠品，租赁期满归承租人所有的租赁货物，进料加工进出口货物，边境地方贸易及边境地区小额贸易进出口货物(边民互市贸易除外)，中外合资企业、中外合作经营企业、外商独资经营企业进出口货物和公用物品，到、离岸价格在规定限额以上的进出口货样和广告品(无商业价值、无使用价值和免费提供出口的除外)，从保税仓库提取在中国境内销售的进口货物，以及其他进出口货物。该指标可以观察一个国家在对外贸易方面的总规模。我国规定出口货物按离岸价格统计，进口货物按到岸价格统计。

14. 商品经营单位所在地进、出口额　指在所在地海关注册登记的有进出口经营权的企业实际进、出口额。

15. 城镇家庭可支配收入　指家庭成员得到可用于最终消费支出和其他非义务性支出以及储蓄的总和，即居民家庭可以用来自由支配的收入。它是家庭总收入扣除交纳的所得税、个人交纳的社会保障支出以及记账补贴后的收入。计算公式为

可支配收入=家庭总收入-交纳所得税-个人交纳的社会保障支出-记账补贴

16. 城镇家庭消费性支出　指家庭用于日常生活的支出，包括食品、衣着、家庭设备用品及服

务、医疗保健、交通和通信、娱乐教育文化服务、居住、杂项商品和服务等八大类支出。

17. 总收入 指调查期内农村住户和住户成员从各种来源渠道得到的收入总和。按收入的性质划分为工资性收入、家庭经营收入、财产性收入和转移性收入。

18. 纯收入 指农村住户当年从各个来源得到的总收入相应地扣除所发生的费用后的收入总和。计算方法

纯收入=总收入-税费支出-家庭经营费用支出-生产性固定资产折旧-赠送农村亲友支出

纯收入主要用于再生产投入和当年生活消费支出，也可用于储蓄和各种非义务性支出。“农民人均纯收入”按人口平均的纯收入水平，反映的是一个地区或一个农户农村居民的平均收入水平。

19. 人口数 指一定时点、一定地区范围内有生命的个人总和。

年度统计的年末人口数指每年 12 月 31 日 24 时的人口数。年度统计的全国人口总数内未包括香港、澳门特别行政区和台湾省以及海外华侨人数。

20. 城镇人口 城镇人口是指居住在城镇范围内的全部常住人口

21. 就业人员 指在 16 周岁及以上，从事一定社会劳动并取得劳动报酬或经营收入的人员。这一指标反映了一定时期内全部劳动力资源的实际利用情况，是研究我国基本国情国力的重要指标。

Explanatory Notes on Main Statistical Indicators

1. Gross Domestic Product (GDP) refers to the final result of the primary distribution of the income created by all the resident units of a country (or a region) during a certain period of time. Gross domestic product is expressed in three different forms, i.e. value, income, and products respectively. The form of value refers to the total value of all products and services produced by all resident units during a certain period of time minus total value of intermediate input of materials and services of the nature of non-fixed assets or the summation of the value added of all resident units: the form of income includes all the income created by all resident units and distributed primarily to all resident and non-resident units; the form of products refers to the summation of the value of the products and services finally used and the net export value of products and services by all resident units during a given period of time. In the practice of national accounting, gross domestic product is calculated with three approaches, i.e. production approach, income approach, and expenditure approach, which reflect the gross domestic product and its composition from different aspects.

2. Three Industries Industrial structure is classified according to the sequence of historical development of social productive activities. Primary industry refers to the extraction of natural resources; secondary industry involves processing of primary products; and tertiary industry provides services of various kinds for production and consumption. The above classification is universal in the world although it varies to some extent from country to country. The three industries in China are divided as follows.

Primary industry: refers to farming, forestry, sideline production and fishery.

Secondary industry: refers to such industries as mining, manufacturing, production and supply of electric power, fuel gas and water, and construction.

Tertiary industry: refers to all the other industries not included in the primary or secondary industries. It includes such industries as communications and transportation, storage and postal service; information transmission, computer service and software; wholesale and retailing; accommodation and catering; financial service; real estate; charter business and commercial affairs service; scientific research, technological service and geological survey; water conservancy, environmental and other public facilities management; residents service and other service trades; education, health, social security and social welfare; culture, sports and entertainment business as well as public administration and social organizations, and international organizations.

3. Added Value refers to the final result of production operation and labor activities of all trades and professions, which is calculated by using the methods of production and income.

Production Method: refers to the method whereby to get the newly added value by proceeding from the gross product of goods and service activities occurring in the course of production and then rejecting the value of intermediate product input in the course of production.

Income Method: is also called the distribution method. The calculation of the gross domestic product (GDP) by the income method is the accounting of the result of productive activities of permanent units from the angle of the income created in the course of production. According to this method, the added value is composed of the payment for laborers, depreciation for fixed assets, net tax on production and business surplus.

4. Total Population by the End of the Year refers to the sum of living individuals within a particular range of area at 24:00 on December 31 of each year.

5. Government Revenue refers to the income obtained by the government finance through participating in the distribution of social products. It is the financial guarantee to ensure government functioning. The contents of government revenue have changed several times. Now it includes the following main items:

(1) Various tax revenues, including value added tax, business tax, consumption tax, land value-added tax, tax on city maintenance and construction, resources tax, tax on use of urban land, enterprise income tax, personal income tax, tariff, stamp tax on security transactions, tax on purchase of motor vehicles, tax on agriculture and animal husbandry and tax on occupancy of cultivated land, etc.

(2) Special revenues, including revenues from the fee on sewage treatment, fee on urban water resources, fee for the compensation of mineral resources and extra-charges for education, etc.

(3) Other revenues, including revenues from interest, repayment of capital construction loan, capital construction projects, and donations and grants.

(4) Subsidies for the losses of State-owned enterprises. This is an item of negative revenue, counteracting revenues and consisting of subsidies to industrial, commercial and grain purchasing and supply enterprises.

6. Government Expenditure refers to the distribution and use of the funds which the government finance has raised, so as to meet the needs of economic construction and various causes.

7. Expenditure for capital construction It refers to the non-gratuitous use of, appropriation of funds for and capital outlay on capital construction in the area of capital construction. It also covers the loans on capital construction approved by the government for special purposes or policy purposes and the expenditure with discount paid in an overall way within the amount of the funds appropriated to the departments for capital construction.

8. Regular Institutions of Higher Learning refer to educational establishments set up according to the government evaluation and approval procedures, enrolling graduates from senior secondary schools and providing higher education courses and training for senior professionals. They include full-time universities, colleges, institutions of higher professional education, institutions of higher vocational education and others.

Universities and colleges primarily provide undergraduate courses; institutions of higher professional education and institutions of higher vocational education primarily provide professional trainings; and others refer to educational establishments, which are responsible for enrolling higher education students under the State Plan but not enumerated in the total number of schools, including: branch schools of universities and colleges, and universities and colleges that have been approved and under plan for construction.

9. Health Care Institutions include: medical institutions, disease prevention and control centres (epidemic prevention stations), blood gathering and supplying institutions, health supervision and inspection (check up) institutions, medicinal scientific research and on-job training institutions, health education centres and so on.

10. Medical Organizations include: hospitals, health service centres (stations) in communities, sanatoria, health centres, out-patient clinics, clinics (health stations and infirmaries), maternity and child care agencies (centres and stations), special disease prevention and curing agencies (centres and stations), first aid centres (stations) and clinical inspection centres. Medical organizations are grouped by two types: profit-making and non-profit-making medical organizations.

11. Gross Amount of Water Used refers to gross water use distributed to users, including loss during transportation, broken down into use by agriculture, industry, living consumption and ecological protection.

12. Total Investment in Fixed Assets in the Whole Country refers to the volume of activities in construction and purchases of fixed assets of the whole country and related fees, expressed in monetary terms during the reference period. It is a comprehensive indicator which shows the size, structure and growth of the investment in fixed assets, providing a basis for observing the progress of construction projects and evaluating results of investment. Total investment in fixed assets in the whole country includes, by type of ownership, the investment by State-owned units, collective-owned units, individuals, joint ownership units, share-holding units, as well as investments by entrepreneurs from foreign countries and from Hong Kong, Macao and Taiwan, and by other units.

13. Total Imports and Exports at Customs refer to the real value of commodities imported and exported across the border of China. They include the actual imports and exports through foreign trade, imported and exported goods under the processing and assembling trades and materials, supplies and gifts as aid given gratis between governments and by the United Nations and other international organizations, and contributions donated by overseas Chinese compatriots in Hong Kong and Macao and Chinese with foreign citizenship, leasing commodities owned by tenant at the expiration of leasing period, the imported and exported commodities processed with imported materials, commodities trading in border areas (excluding mutual exchange goods), the imported and exported commodities and articles for public use of the Sino-foreign joint ventures, cooperative enterprises and ventures with sole foreign investment. Also included is the import or export of samples and advertising goods for which the CIF or FOB value is beyond the permitted ceiling (excluding goods of no trading or use value and free commodities for export), imported goods sold in China from bonded warehouses and other imported or exported goods. The indicator of the total imports and exports at customs can be used to observe the total size of external trade in a country. In accordance with the stipulation of the Chinese government, imports are calculated at CIF, while exports are calculated at FOB.

14. Import-Export Value by Location of China's Foreign Trade Managing Units refers to actual value of imports and exports carried out by corporations which have been registered by the local customs house and are vested with right to run import export business.

15. Disposable Income of Urban Households refers to the actual income at the disposal of members of the households which can be used for final consumption, other non-compulsory expenditure and savings. This equals to total income minus income tax, personal contribution to social security and subsidy for keeping diaries in being a sample household. The following formula is used:

Disposable income = total household income − income tax − personal contribution to social security-subsidy for keeping diaries for a sampled household

16. Consumption Expenditure of Urban Households refers to total expenditure of households for consumption in daily life, including expenditure on the eight categories of food; clothing; household appliances and services; health care and medical services; transport and communications; recreation, education and cultural services; housing; and miscellaneous goods and services.

17. Total Income refers to the sum of income earned from various sources by the rural households and their members during the reference period, and is classified as income from wages and salaries, household operations, properties and transfers.

18. Net Income refers to the total income of rural households from all sources minus all corresponding expenses. The formula for calculation is as follows:

Net income = total income − taxes and fees paid − household operation expenses − taxes and fees − depreciation of fixed assets for production − gifts to non-rural relatives

Net income is mainly used as input for reinvestment in production and as consumption expenditure of the year, and also used for savings and non-compulsory expenses of various forms. "Per capita net income of farmers" is the level of net income averaged by population, reflecting the average income level of rural households in a given area.

19. Total Population refers to the total number of people alive at a certain point of time within a given area.

The annual statistics on total population is taken at midnight, the 3lst of December, not including residents in Taiwan province, Hong Kong and Macao and overseas Chinese.

20. Urban Population refers to all people residing in cities and towns.

21. Employed Persons refers to persons aged 16 and over who are engaged in gainful employment and thus receive remuneration payment or earn business income. This indicator reflects the actual utilization of total labour force during a certain period of time and is often used for the research on China's economic situation and national power.

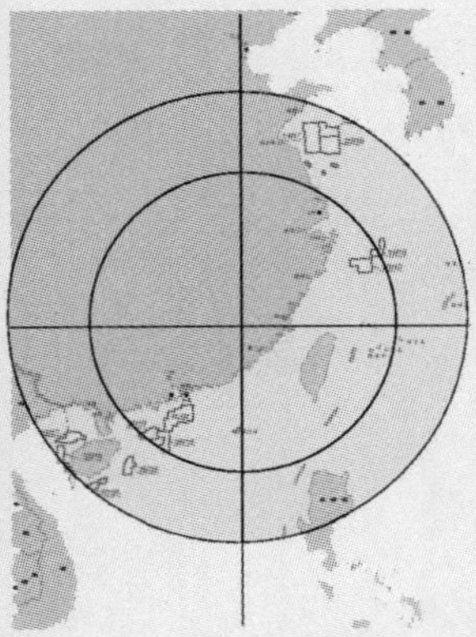

11

部分世界海洋经济统计资料
Part of the World's Marine Economic Statistics Data

资料来源：

11-1 世界海洋面积（2010年）
World Ocean Area, 2010

区 域 Region	海洋面积 （平方千米） Ocean Area (km^2)	占世界海洋面积的比重（%） Proportion in the World Ocean Area (%)	占地球表面面积的比重（%） Proportion in the Earth's Surface Area (%)
合 计 Total	**361 000 000**	**100.0**	**70.8**
太 平 洋 Pacific Ocean	178 334 000	49.4	35.0
大 西 洋 Atlantic Ocean	91 694 000	25.4	18.0
印 度 洋 Indian Ocean	76 171 000	21.1	14.9
北 冰 洋 Arctic Ocean	14 801 000	4.1	2.9

资料来源：《2011国际统计年鉴》。
Data Source: *International Statistical Yearbook 2011*.

11-2 主要沿海国家（地区）海岸线长度
Length of Coastline of Major Coastal Countries (Regions)

单位：千米 (km)

国家和地区 Country and Region	海岸线长度 Length of Coastline
中国 China	32 000
美国 United States	22 680
日本 Japan	30 000
德国 Germany	1 300
英国 United Kingdom	11 450
法国 France	3 000
意大利 Italy	7 000
加拿大 Canada	20 000
澳大利亚 Australia	20 125
俄罗斯联邦 Russian Fed.	34 000
波兰 Poland	491
印度 India	6 083
印度尼西亚 Indonesia	35 000
菲律宾 Philippines	18 533
泰国 Thailand	3 219
马来西亚 Malaysia	4 675
新加坡 Singapore	193
缅甸 Myanmar	3 060
孟加拉国 Bangladesh	580
土耳其 Turkey	7 200
韩国 Korea, Rep.	2 413
埃及 Egypt	2 450
墨西哥 Mexico	9 330
巴西 Brazil	7 400
阿根廷 Argentina	4 989

11-3 世界主要沿海国家（地区）国土面积和人口
Surface Area and Population of Major Coastal Countries (Regions)

国家和地区 Country and Region	国土面积*（万平方千米） Area of Territory (10 000 km^2)	年中人口（万人） Mid-year Population (10 000 persons)	人口密度*（人/平方千米） Population Density (people per km^2)
世界总计 World Total	**13 432.5**	**720 774**	**56**
中国 China	960.0	136 427	145
文莱 Brunei Darsm	0.6	42	79
柬埔寨 Cambodia	18.1	1 541	86
印度 India	328.7	126 740	421
印度尼西亚 Indonesia	191.1	25 281	138
日本 Japan	37.8	12 713	349
韩国 Korea, Rep.	10.0	5 042	516
马来西亚 Malaysia	33.1	3 019	91
缅甸 Myanmar	67.7	5 372	82
菲律宾 Philippines	30.0	10 010	330
新加坡 Singapore	0.1	547	7 713
泰国 Thailand	51.3	6 722	131
越南 Viet Nam	33.1	9 073	289
埃及 Egypt	100.2	8 339	82
南非 South Africa	121.9	5 400	44
加拿大 Canada	998.5	3 554	4
墨西哥 Mexico	196.4	12 380	63
美国 United States	983.2	31 886	35
阿根廷 Argentina	278.0	4 180	15
巴西 Brazil	851.6	20 203	24
法国 France	54.9	6 620	120
德国 Germany	35.7	8 089	231
意大利 Italy	30.1	6 134	205
俄罗斯联邦 Russian Fed.	1 709.8	14 382	9
西班牙 Spain	50.6	4 641	94
土耳其 Turkey	78.4	7 584	97
乌克兰 Ukraine	60.4	4 536	79
英国 United Kingdom	24.4	6 451	265
澳大利亚 Australia	774.1	2 349	3
新西兰 New Zealand	26.8	451	17

资料来源：《2015年中国统计年鉴》。

Data Source: *China Statistical Yearbook 2015*.

*2013年数据。

*Data for 2013.

11-4　主要沿海国家（地区）国内生产总值
Gross Domestic Product of Major Coastal Countries (Regions)

国家和地区 Country and Region	国内生产总值（亿美元） GDP (100 million USD)	人均国内生产总值（美元） GNI per Captia (current USD)	国内生产总值增长率(%) Growth Rate of GDP(%)
中国 China	103 601	7 594	7.3
中国香港 Hong Kong, China	2 909	40 170	2.5
文莱 Brunei Darsm	173	40 776	5.3
柬埔寨 Cambodia	167	1 084	7.0
印度 India	20 669	1 631	7.4
印度尼西亚 Indonesia	8 885	3 515	5.0
日本 Japan	46 015	36 194	-0.1
韩国 Korea, Rep.	14 104	27 971	3.3
马来西亚 Malaysia	3 269	10 830	6.0
菲律宾 Philippines	2 846	2 843	6.1
新加坡 Singapore	3 079	56 287	2.9
泰国 Thailand	3 738	5 561	0.7
越南 Viet Nam	1 862	2 052	6.0
埃及 Egypt	2 865	3 436	2.2
南非 South Africa	3 498	6 478	1.5
加拿大 Canada	17 867	50 271	2.5
墨西哥 Mexico	12 827	10 361	2.1
美国 United States	174 190	54 630	2.4
阿根廷 Argentina	5 402	12 922	0.5
巴西 Brazil	23 461	11 613	0.1
法国 France	28 292	42 736	0.2
德国 Germany	38 526	47 627	1.6
意大利 Italy	21 443	34 960	-0.4
荷兰 Netherlands	8 695	51 590	0.9
波兰 Poland	5 480	14 423	3.4
俄罗斯联邦 Russian Fed.	18 606	12 736	0.6
西班牙 Spain	14 043	30 262	1.4
土耳其 Turkey	7 995	10 543	2.9
乌克兰 Ukraine	1 318	3 083	-6.8
英国 United Kingdom	29 419	45 603	2.6
澳大利亚 Australia	14 538	61 887	2.5

资料来源：世界银行WDI数据库。

Data Source: *World Bank WDI Database.*

11-5 主要沿海国家（地区）国内生产总值产业构成
Industrial Composition of GDP of Major Coastal Countries (Regions) by Industry

国家和地区 Country and Region	国内生产总值产业构成（%） Industrial Structure of GDP(%)		
	第一产业 Primary Industry	第二产业 Secondary Industry	第三产业 Tertiary Industry
世界 World	**3.1** ①	**26.4** ①	**70.5** ①
中国 China	9.2	42.6	48.2
中国香港 Hong Kong,China	0.1 ①	7.2 ①	92.7 ①
文莱 Brunei Darsm	0.7 ①	68.2 ①	31.0 ①
柬埔寨 Cambodia	29.8	27.3	42.9
印度 India	17.0	30.1	53.0
印度尼西亚 Indonesia	13.7	43.0	43.3
日本 Japan	1.2 ①	26.2 ①	72.6 ①
韩国 Korea, Rep.	2.3	38.2	59.4
马来西亚 Malaysia	9.1	40.5	50.4
缅甸 Myanmar			
菲律宾 Philippines	11.3	31.2	57.5
新加坡 Singapore		25.1 ①	74.9 ①
斯里兰卡 Sri Lanka	9.9	33.8	56.3
泰国 Thailand	11.6	42.0	46.3
越南 Viet Nam	18.1	38.5	43.4
埃及 Egypt	14.5	40.0	45.6
南非 South Africa	2.5	29.5	68.1
加拿大 Canada			
墨西哥 Mexico	3.5	33.8	62.7
美国 United States	1.5 ①	20.5 ①	78.1 ①
阿根廷 Argentina	8.2	28.8	63.0
巴西 Brazil	5.6	23.4	71.0
法国 France	1.7	19.4	78.9
德国 Germany	0.8	30.7	68.6
意大利 Italy	2.2	23.4	74.4
荷兰 Netherlands	2.0	21.4	76.6
俄罗斯联邦 Russian Fed.	4.0 ①	36.3 ①	59.8 ①
西班牙 Spain	2.5	23.1	74.4
土耳其 Turkey	8.0	27.1	64.9
英国 United Kingdom	0.6	19.8	79.6
澳大利亚 Australia	2.5	27.1	70.4
新西兰 New Zealand	6.9 ②	23.4 ②	69.7 ②

注：①2013年数据。②2011年数据。

资料来源：世界银行WDI数据库。

Note: ①Data for 2013. ②Data for 2011.

Data Source: *World Bank WDI Database.*

11-6 主要沿海国家（地区）就业人数
Employment in the Major Coastal Countries (Regions)

单位：万人　　(10 000 persons)

国家和地区 Country and Region	2000	2005	2007	2008	2009	2010	2011	2012	2013
中国 China	72 085	74 647	76 990	75 564	75 828	76 105	76 420	76 704	76 977
中国香港 Hong Kong,China	321	334	319	353	347	350	358	366	373
印度 India	33 020	37 199				37 429		39 267	
印度尼西亚 Indonesia	8 984	9 396	9 993	10 255	10 487	10 821	10 967	11 301	11 276
以色列 Israel	222	249	268	278	284	294	302	336	345
日本 Japan	6 446	6 356	6 412	6 385	6 282	6 256	6 289	6 270	6 311
韩国 Korea, Rep.	2 116	2 286	2 343	2 358	2 351	2 383	2 424	2 468	2 507
马来西亚 Malaysia	932	1 005	1 102	1 066	1 090	1 113	1 228	1 272	
菲律宾 Philippines	2 745	3 231	3 356	3 409	3 506	3 604	3 719	3 762	3 792
新加坡 Singapore	209	165	180	185	187		200	204	
斯里兰卡 Sri Lanka	631	752	704	765	760	771	820	813	848
泰国 Thailand	3 300	3 630	3 625	3 784	3 771	3 804	3 846	3 894	
越南 Viet Nam	3 837	4 253					5 088	5 169	
埃及 Egypt	1 720	1 934	2 153	2 251	2 298	2 383	2 335	2 360	2 397
南非 South Africa	1 224	1 230	1 347	1 371	1 346	1 306	1 326	1 352	
加拿大 Canada	1 476	1 617	1 681	1 713	1 681	1 704	1 731	1 751	
墨西哥 Mexico	3 950	4 079	4 306	4 387	4 291	4 387	4 607	4 900	
美国 United States	13 521	14 173	14 605	14 536	13 988	13 906			
阿根廷 Argentina	826	964	1 016	1 028	1 034	1 053	1 077	1 084	
巴西 Brazil	6 563	8 719	2 044	9 239	9 269	2 202	9 247	2 296	2 312
委内瑞拉 Venezuela	896	1 073	1 132	1 186	1 194	1 207		1 239	
法国 France	2 312	2 495	2 558	2 590	2 564	2 569	2 578	2 580	2 575
德国 Germany	3 632	3 636	3 979	3 854	3 847	3 874	3 974	4 006	3 953
意大利 Italy	2 093	2 256	2 322	2 340	2 303	2 287	2 297	2 290	2 219
荷兰 Netherlands	786	811	731	859	860	837	837	842	
波兰 Poland	1 453	1 412	1 524	1 580	1 587	1 596	1 613	1 559	
俄罗斯联邦 Russian Fed.	6 507	6 817	7 055	7 097	6 928	6 980	7 086	7 155	
西班牙 Spain	1 544	1 897	2 036	2 026	1 889	1 846	1 810	1 728	1 714
土耳其 Turkey	2 158	2 205	2 079	2 119	2 128	2 259	2 410	2 482	
乌克兰 Ukraine	2 127	2 068	2 091	2 097	2 019	2 027	2 032	2 035	
英国 United Kingdom	2 740	2 867	2 923	2 936	2 892	2 894	2 908	2 943	2 995
澳大利亚 Australia	895	997	1 058	1 074	1 092	1 121	1 136	1 148	
新西兰 New Zealand	181	207	217	219	216	218	222	222	

资料来源：联合国ILO数据库。

Data Source: *ILO Database.*

11-7　主要沿海国家（地区）鱼类产量（2013年）
Fish Yields of Major Coastal Countries (Regions), 2013

单位：万吨　　(10 000 tons)

国家和地区 Country and Region	鱼类产量 Output of Total Fishes	海域鱼类产量 Ocean Area
中国 China	3 740.2	1 092.4
印度 India	826.5	293.6
印度尼西亚 Indonesia	865.3	580.6
缅甸 Myanmar	460.5	243.3
俄罗斯联邦 Russian Fed.	432.2	393.2
美国 United States	437.0	414.7
越南 Viet Nam	451.6	196.3
日本 Japan	314.3	309.3
孟加拉国 Bangladesh	318.3	61.4
菲律宾 Philippines	282.6	237.9
泰国 Thailand	197.5	129.6
马来西亚 Malaysia	143.8	130.1
巴西 Brazil	108.0	45.8
墨西哥 Mexico	141.0	125.8
埃及 Egypt	142.5	9.1
韩国 Korea, Rep.	119.5	117.4
西班牙 Spain	100.9	98.8
尼日利亚 Nigeria	96.1	34.3
南非 South Africa	40.9	40.6
英国 United Kingdom	64.4	63.0
柬埔寨 Cambodia	70.2	8.8
土耳其 Turkey	56.1	40.6
阿根廷 Argentina	53.5	51.9
加拿大 Canada	53.6	49.9
斯里兰卡 Sri Lanka	49.5	40.2
新西兰 New Zealand	42.0	41.9
法国 France	45.2	41.2
荷兰 Netherlands	30.8	30.3

资料来源：联合国FAO数据库。
Data Source: *FAO Database*.

11-8 国家保护区面积和鱼类濒危物种
Area of National Nature Reserves and Endangered Species of Fish

国家和地区 Country and Region	国家保护区[①] National Protected[①]		鱼类濒危物种（种） Endangered Species of Fish (number)
	陆地保护区面积占陆地面积比重 Terrestrial Protected Areas (% of Total Land Area)	海洋保护区面积占领海面积比重 Marine Protected Areas (% of Territorial Waters)	
中国 China	16.7	1.6	122
中国香港 Hong Kong, China	41.9		13
孟加拉国 Bangladesh	4.7	2.5	18
文莱 Brunei Darsm	44.0	1.4	7
柬埔寨 Cambodia	26.2	0.5	40
印度 India	5.2	1.6	213
印度尼西亚 Indonesia	14.7	5.8	145
伊朗 Iran	7.2	2.2	37
日本 Japan	16.5	5.6	67
韩国 Korea, Rep.	6.3	3.9	19
马来西亚 Malaysia	18.4	2.3	73
菲律宾 Philippines	10.9	2.5	74
新加坡 Singapore	5.4	1.4	25
斯里兰卡 Sri Lanka	22.0	1.3	43
泰国 Thailand	18.8	5.1	96
越南 Viet Nam	6.5	1.7	73
埃及 Egypt	11.2	13.1	40
尼日利亚 Nigeria	14.1	0.2	60
南非 South Africa	6.2	12.8	87
加拿大 Canada	8.6	1.3	36
墨西哥 Mexico	12.9	18.9	154
美国 United States	13.8	30.4	236
阿根廷 Argentina	6.9	1.6	37
巴西 Brazil	26.3	16.3	84
委内瑞拉 Venezuela	53.0	16.0	37
法国 France	24.7	58.5	46
德国 Germany	48.0	64.5	23
意大利 Italy	21.6	19.9	47
荷兰 Netherlands	19.5	61.8	13
波兰 Poland	34.2	52.8	7
俄罗斯联邦 Russian Fed.	11.3	11.6	36
西班牙 Spain	29.0	9.4	70
土耳其 Turkey	2.1	2.7	126
乌克兰 Ukraine	4.0	10.7	21
英国 United Kingdom	27.9	16.6	43
澳大利亚 Australia	12.8	33.2	106
新西兰 New Zealand	27.3	12.4	23

注：①2012年数据。
Note: ①Data refer to 2012.
资料来源：世界银行WDI数据库。
Data Source: *World Bank WDI Database.*

11-9 主要沿海国家（地区）风力发电量
Wind-Power Capacities of Major Coastal Countries (Regions)

单位：百万千瓦·时

(million kilowatt hours)

国家和地区 Country and Region	2006	2007	2008	2009	2010	2011
中国 China			13 079	46 096	44 622	70 331
伊朗 Iran		143	196	224		
以色列 Israel	2	1	9	9	8	
日本 Japan	1 753	2 624	2 623	2 949	3 962	7
韩国 Korea, Rep	239	376	436	685	817	863
马来西亚 Malaysia						
巴基斯坦 Pakistan						
菲律宾 Philippines	55	59	61	64	62	88
新加坡 Singapore						
斯里兰卡 Sri Lanka	2	2	3	3	53	92
泰国 Thailand	1				3	3
越南 Viet Nam						
土耳其 Turkey						
埃及 Egypt	616	831	931	1 133	1 498	1 525
尼日利亚 Nigeria						
南非 South Africa	32	32	32	32	32	103
加拿大 Canada	2 500	3 024	3 819	4 573	9 557	10 187
墨西哥 Mexico	59	262	269	596	1 239	1 648
美国 United States	26 676	34 603	55 696	74 226	95 148	120 854
阿根廷 Argentina	70	61	42	36	25	26
巴西 Brazil						
委内瑞拉 Venezuela						
法国 France	2 150	4 052	5 689	7 891	9 969	12 052
德国 Germany	30 710	39 713	40 574	38 639	37 793	48 883
意大利 Italy	2 971	4 034	4 861	6 543	9 126	9 856
荷兰 Netherlands	2 733	3 438	4 260	4 581	3 993	5 100
波兰 Poland	256	522	837	1 077	1 664	3 205
俄罗斯联邦 Russian Fed.	5	7	5	4	4	5
西班牙 Spain	23 040	27 509	32 203	37 773	44 165	42 918
土耳其 Turkey	127	355	847	1 495	2 916	4 723
乌克兰 Ukraine	35	45	45	43	50	89
英国 United Kingdom	4 225	5 274	7 097	9 304	10 183	15 509
澳大利亚 Australia	1 691	2 611	3 941	3 806	4 798	5 807
新西兰 New Zealand	623	937	1 057	1 471	1 634	1 952

资料来源：联合国ESD数据库。

Data Source: UN ESD Database.

11-10 主要沿海国家（地区）捕捞产量
Fishing Yields in the Major Coastal Countries(Regions)

单位：吨 (t)

国家和地区 Country and Region	2012	2013
世界总计 World Total	**91 305 829**	**92 572 586**
中国 China	16 167 443	16 274 926
秘鲁 Peru	4 849 211	5 854 233
印度尼西亚 Indonesia	5 813 800	6 101 725
美国 United States	5 128 381	5 230 874
印度 India	4 872 129	4 645 182
俄罗斯联邦 Russian Fed.	4 331 398	4 345 868
日本 Japan	3 650 950 *	3 656 854 *
缅甸 Myanmar	3 579 250	3 786 840
智利 Chile	2 572 876	1 770 945
越南 Viet Nam	2 705 400	2 803 800
菲律宾 Philippines	2 322 974	2 331 721
挪威 Norway	2 150 555	2 074 363
泰国 Thailand	1 719 628	1 843 747
韩国 Korea, Rep.	1 670 122	1 597 874
孟加拉国 Bangladesh	1 535 715	1 550 446
墨西哥 Mexico	1 575 409	1 626 869
马来西亚 Malaysia	1 477 281	1 488 705
冰岛 Iceland	1 358 596	1 366 675
西班牙 Spain	931 583	1 034 173
摩洛哥 Morocco	1 161 511	1 253 299
中国台湾 Taiwan China	908 088	925 268
加拿大 Canada	818 820	852 082
巴西 Brazil	842 987	765 287
阿根廷 Argentina	738 060	870 083
南非 South Africa	702 761	412 510
尼日利亚 Nigeria	668 754	721 355
英国 United Kingdom	631 458	632 345
柬埔寨 Cambodia	566 695	639 468
伊朗 Iran	542 378	559 631
丹麦 Denmark	502 790	668 480
斯里兰卡 Sri Lanka	473 832	495 573
新西兰 New Zealand	440 683	442 448
土耳其 Turkey	432 444	374 128
法国 France	456 796	493 256
埃及 Egypt	354 237	356 857
荷兰 Netherlands	347 264	327 437
委内瑞拉 Venezuela	213 072	200 011 *
德国 Germany	207 500	229 904
意大利 Italy	200 816	

注：捕捞品种包括鱼类、甲壳类、软体类等水生动物。*为联合国粮农组织估算值。

资料来源：《渔业和水产养殖统计年鉴》，联合国粮农组织，2013年 (表11-11同) 。

Note: The fished species include fish, crustacea, mollusc and aquatic animals.

* It is estimated by FAO from available sources of information or calculation.

Source: *Fishery and AquacultureStatistical Yearbook*, FAO, 2013 (Same as Table 11-11).

11-11 主要沿海国家水产养殖产量
Aquaculture Production in the Major Coastal Countries (Regions)

单位：吨 (t)

国家和地区 Country and Region	2012	2013
世界总计 World Total	**66 477 284**	**70 189 830**
中国 China	41 108 306	43 549 738
印度 India	4 209 478	4 549 607
越南 Viet Nam	3 085 500	3 207 200
印度尼西亚 Indonesia	3 067 660	3 819 732
孟加拉国 Bangladesh	1 726 066	1 859 808
挪威 Norway	1 321 119	1 247 865
泰国 Thailand	1 272 100	1 056 944
智利 Chile	1 071 421	1 033 206
埃及 Egypt	1 017 738	1 097 544
缅甸 Myanmar	885 169	929 180
菲律宾 Philippines	790 894	815 008
巴西 Brazil	480 150	473 429
日本 Japan	633 047	608 800
韩国 Korea, Rep.	486 900	402 141
美国 United States	419 974	441 098
中国台湾 Taiwan China	344 404	344 453
伊朗 Iran	296 575	325 325
马来西亚 Malaysia	303 386	261 274
西班牙 Spain	264 161	223 707
尼日利亚 Nigeria	253 898	278 706
土耳其 Turkey	212 805	233 864
法国 France	204 860	201 860
英国 United Kingdom	203 037	194 630
加拿大 Canada	173 252	172 097
意大利 Italy	162 618	162 620
俄罗斯联邦 Russian Fed.	144 871	154 898
墨西哥 Mexico	143 747	168 792

注：养殖品种包括鱼类、甲壳类、软体类等水生动物。

Note: The cultivated varieties includes fish, crustacea, mollusc and other aquatic animals.

11-12 主要沿海国家（地区）石油主要指标（2013年）
Main Oil Indicators of Major Coastal Countries (Regions), 2013

国家和地区 Country and Region	原油探明储量（亿桶） Crude Oil Proved Reserves (100 million Barrels)	石油存量（万桶） Total Petroleum Stocks (10 000 Barrels)
世界 World	**16 489**	**349 339**
中国 China	240	
孟加拉国 Bangladesh		
文莱 Brunei Darsm		
印度 India	55	
印度尼西亚 Indonesia	40	
伊朗 Iran	1 550	
以色列 Israel		
日本 Japan		57 500
韩国 Korea, Rep.		17 800
马来西亚 Malaysia	40	
缅甸 Myanmar		
巴基斯坦 Pakistan		
菲律宾 Philippines		
泰国 Thailand		
越南 Viet Nam	44	
埃及 Egypt	44	
尼日利亚 Nigeria	370	
南非 South Africa		
加拿大 Canada	1 730	17 000
墨西哥 Mexico	100	4 900
美国 United States	330	176 100
阿根廷 Argentina	28	
巴西 Brazil	130	
委内瑞拉 Venezuela	2 980	
法国 France		16 700
德国 Germany		29 000
意大利 Italy		12 500
荷兰 Netherlands		11 600
波兰 Poland		6 000
俄罗斯联邦 Russian Fed.	800	
西班牙 Spain		11 200
土耳其 Turkey		6 200
乌克兰 Ukraine		
英国 United Kingdom	31	7 800
澳大利亚 Australia		3 700
新西兰 New Zealand		830

资料来源：美国能源署。
Data Source: U. S.Energy Information Administration.

11-13 万美元国内生产总值能耗
Energy Use per Ten Thousand USD of GDP

单位：吨标准油/万美元 (ton of oil equivalent per 10 000 USD)

国家和地区 Country and Region	2011年不变价， PPP (Constant 2011，PPP)		
	2010	2011	2012
世界 World	**1.38**	**1.36**	**1.34**
中国 China	2.00	1.99	1.95
中国香港 Hong Kong,China	0.42	0.43	0.41
孟加拉国 Bangladesh	0.82	0.80	0.79
文莱 Brunei Darsm	1.15	1.32	1.31
柬埔寨 Cambodia	1.39	1.38	1.32
印度 India	1.32	1.29	1.28
印度尼西亚 Indonesia	1.02	0.95	0.93
伊朗 Iran	1.64	1.62	1.79
以色列 Israel	1.03	0.99	1.01
日本 Japan	1.13	1.05	1.01
韩国 Korea, Rep.	1.66	1.67	1.65
马来西亚 Malaysia	1.31	1.30	1.27
巴基斯坦 Pakistan	1.16	1.13	1.10
菲律宾 Philippines	0.77	0.74	0.73
新加坡 Singapore	0.69	0.67	0.62
斯里兰卡 Sri Lanka	0.62	0.61	0.63
泰国 Thailand	1.38	1.40	1.39
越南 Viet Nam	1.51	1.47	1.49
埃及 Egypt	0.88	0.90	0.91
尼日利亚 Nigeria	1.47	1.48	1.50
南非 South Africa	2.32	2.24	2.16
加拿大 Canada	1.81	1.77	1.73
墨西哥 Mexico	0.97	0.97	0.96
美国 United States	1.45	1.41	1.35
阿根廷 Argentina			
巴西 Brazil	0.93	0.91	0.93
委内瑞拉 Venezuela	1.57	1.41	1.45
法国 France	1.09	1.03	1.03
德国 Germany	0.99	0.90	0.90
意大利 Italy	0.80	0.79	0.77
荷兰 Netherlands	1.10	1.00	1.03
波兰 Poland	1.24	1.19	1.13
俄罗斯联邦 Russian Fed.	2.27	2.29	2.27
西班牙 Spain	0.83	0.82	0.84
土耳其 Turkey	0.88	0.86	0.87
乌克兰 Ukraine	3.68	3.34	3.23
英国 United Kingdom	0.89	0.81	0.83
澳大利亚 Australia	1.35	1.32	1.33
新西兰 New Zealand	1.32	1.29	1.31

资料来源：世界银行WDI数据库。

Data Source: *World Bank WDI Database.*

11-14 主要沿海国家（地区）国际海运装货量和卸货量
International Maritime Freight Loaded and Unloaded in the Major Coastal Countries (Regions)

单位：万吨 (10 000 t)

国家和地区 Country and Region	国际海运装货量 International Ocean Shipping Loading Capacity			国际海运卸货量 International Ocean Shipping Unloading Capacity		
	2000	2005	2014	2000	2005	2014
中国香港 Hong Kong,China	6 770	8 918	11 342	10 693	14 095	18 508
孟加拉国 Bangladesh	89	79	581	1 408		4 820
文莱 Brunei Darsm	10	8		102	168	
印度尼西亚 Indonesia	14 153	27 372		4 504	8 479	
伊朗 Iran	3 065	3 446		4 486	6 073	
以色列 Israel	1 387	1 663	1 999	2 920	2 107	2 825
日本 Japan	13 010			80 654		
韩国 Korea Rep.	15 078	24 250		41 882	51 245	
马来西亚 Malaysia	5 483	7 940	13 253 ①	6 922	10 391	15 770 ①
巴基斯坦 Pakistan	617	1 126		3 080	3 955	
新加坡 Singapore	32 618	42 266				
斯里兰卡 Sri Lanka	919	1 315				
埃及 Egypt		2 123			4 241	
南非 South Africa	2 598					
美国 United States	34 334	40 534		83 352	94 160	
阿根廷 Argentina	1 550					2 742
法国 France	6 810	10 066	10 354	20 273	22 710	18 888
德国 Germany	8 602	10 832	12 107	14 725	16 866	17 479
荷兰 Netherlands	9 940	12 250		32 507	36 424	
波兰 Poland	3 152	3 835	2 921	1 582	1 642	3 856
俄罗斯联邦 Russian Fed.	828	910		84	74	
西班牙 Spain	5 627	8 195		19 343	26 164	
乌克兰 Ukraine	4 271	7 070	10 114	684	1 333	1 610
澳大利亚 Australia	48 750	62 401	130 394	5 418	6 989	9 893
新西兰 New Zealand	2 214	2 167	3 779	1 379	1 844	2 188

注：①2013年数据。

资料来源：联合国统计月报数据库。

Note: ①Data for 2013.

Data Source: *UN Monthly Bulletin of Statistics Database.*

11-15 世界主要外贸货物海运量及构成*（2013年）
World Major Maritime Freight Traffic in Foreign Trade*, 2013

品 种 Sort	海运量（百万吨） Freight Traffic (million tons)	
	2013	所占比例（%） Percentage
合 计 **Total**	**9 932**	**100.0**
原 油 Crude Oil	1 886	19.0
成品油 Refined Oil	965	9.7
燃 气 Fuel gas	282	2.8
铁矿石 Ironstone	1 197	12.1
煤 炭 Coal	1 116	11.2
谷 物 Corn	370	3.7
其他货物 Others	4 116	41.4

注：*为估计数。
资料来源：Shipping Statistics and Market Review, January/February 2014, ISL
Note: *Estimated data.
Data Source: *Shipping Statistics and Market Review*, January/February 2014, ISL.

11-16 主要沿海国家（地区）国际旅游人数
Number of International Tourists of Major Coastal Countries (Regions)

单位：万人 (10 000 persons)

国家和地区 Country and Region	入境（过夜）旅游人数 Number of Inbound Tourists (Overnight)			出境旅游人数 Number of Outbound Tourists		
	2000	2010	2013	2000	2010	2013
世界总计 World Total	**70 648**	**97 989**	**112 320**	**85 035**	**109 870**	**121 497**
中国 China	3 123	5 566	5 569	1 047	5 739	9 819
中国香港 Hong Kong, China	881	2 009	2 566	461	8 444	8 441
中国澳门 Macao, China	520	1 193	1 427	14	75	145
孟加拉国 Bangladesh	20	30	15	113		
文莱 Brunei Darsm	98	21	23			
柬埔寨 Cambodia	47	251	421	4	51	87
印度 India	265	578	697	442	1 299	1 663
印度尼西亚 Indonesia	506	700	880	221	624	797
伊朗 Iran	134	294	477	229		
以色列 Israel	242	280	296	353	427	476
日本 Japan	476	861	1 036	1 782	1 664	1 747
韩国 Korea, Rep.	532	880	1 218	551	1 249	1 485
马来西亚 Malaysia	1 022	2 458	2 572	3 053		
缅甸 Myanmar	42	79	204			
巴基斯坦 Pakistan	80	91				
菲律宾 Philippines	199	352	468	167		
新加坡 Singapore	606	916	1 190	444	734	865
斯里兰卡 Sri Lanka	40	65	128	52	112	126

11-16 续表 continued

国家和地区 Country and Region	入境（过夜）旅游人数 Number of Inbound Tourists (Overnight)			出境旅游人数 Number of Outbound Tourists		
	2000	2010	2013	2000	2010	2013
泰国 Thailand	958	1 594	2 655	191	545	597
越南 Viet Nam	214	505	757			
埃及 Egypt	512	1 405	917	296	462	578
尼日利亚 Nigeria	81	156	60			
南非 South Africa	587	807	954	383	517	517
加拿大 Canada	1 963	1 622	1 659	1 918	2 868	3 298
墨西哥 Mexico	2 064	2 329	2 415	1 108	1 433	1 591
美国 United States	5 124	6 001	6 977	6 133	6 106	6 157
阿根廷 Argentina	291	533	557	495	531	754
巴西 Brazil	531	516	581	323	650	867
委内瑞拉 Venezuela	47	53	99	95	148	193
法国 France	7 719	7 765	8 473	1 989	2 504	2 624
德国 Germany	1 898	2 688	3 155	7 440		
意大利 Italy	4 118	4 363	4 770	2 199	2 982	2 780
荷兰 Netherlands	1 000	1 088	1 278	1 390	1 837	1 809
波兰 Poland	1 740	1 247	1 580	5 668	4 276	1 005
俄罗斯联邦 Russian Fed.	2 117	2 228	3 079	1 837	3 932	5 407
西班牙 Spain	4 640	5 268	6 066	410	1 238	1 125
土耳其 Turkey	959	3 136	3 780	528	656	753
乌克兰 Ukraine	643	2 120	2 467	1 342	1 718	2 376
英国 United Kingdom	2 321	2 830	3 117	5 684	5 556	5 851
澳大利亚 Australia	493	579	638	350	710	877
新西兰 New Zealand	178	244	263	128	203	219

资料来源：世界银行WDI数据库。

Data source: *WDI database of World Bank.*

11-17 主要沿海国家（地区）国际旅游收入
International Tourism Receipts of Major Coastal Countries (Regions)

单位：亿美元 (10 000 million USD)

国家和地区 Country and Region	国际旅游收入 International Tourism Receipts			
	2010	2011	2012	2013
世界总计 World Total	**11 191**	**12 495**	**12 972**	**13 811**
中国 China	502	533	549	564
中国香港 Hong Kong, China	272	337	380	426
中国澳门 Macao, China	282	390	445	523
孟加拉国 Bangladesh	1	1	1	1
文莱 Brunei Darsm				
柬埔寨 Cambodia	17	18	20	29
印度 India	145	177	183	190
印度尼西亚 Indonesia	76	90	95	103
伊朗 Iran	26	26		
以色列 Israel	58	60	62	65
日本 Japan	154	125	162	169
韩国 Korea, Rep.	144	175	197	193
马来西亚 Malaysia	182	196	203	210
缅甸 Myanmar	1	3		9
巴基斯坦 Pakistan	10	11	10	9
菲律宾 Philippines	34	40	49	56
新加坡 Singapore	142	181	193	191
斯里兰卡 Sri Lanka	10	14	18	25

11-17 续表 continued

国家和地区 Country and Region	国际旅游收入 International Tourism Receipts			
	2010	2011	2012	2013
泰国 Thailand	238	309	377	460
越南 Viet Nam	45	57	68	75
埃及 Egypt	136	93	108	73
尼日利亚 Nigeria	7	7	6	
南非 South Africa	103	107	112	105
加拿大 Canada	184	200	207	177
墨西哥 Mexico	126	125	133	143
美国 United States	1 680	1 845	2 001	2 148
阿根廷 Argentina	56	61	57	50
巴西 Brazil	60	68	69	70
委内瑞拉 Venezuela	8	8	9	
法国 France	561	660	635	661
德国 Germany	491	534	516	552
意大利 Italy	401	454	430	462
荷兰 Netherlands	187	210	205	227
波兰 Poland	100	116	118	125
俄罗斯联邦 Russian Fed.	132	170	179	202
西班牙 Spain	590	677	632	676
土耳其 Turkey	263	301	323	349
乌克兰 Ukraine	47	54	60	60
英国 United Kingdom	408	459	460	494
澳大利亚 Australia	311	342	341	334
新西兰 New Zealand	65	55	55	75

资料来源：世界银行WDI数据库。

Data Source: *World Bank WDI Database.*

11-18 集装箱吞吐量居世界前20位的港口
World Top 20 Seaports in Terms of the Number of Containers Handled

单位：万标准箱 (10 000 TEU)

港口 Seaport	所属国家或地区 Country or Region	吞吐量 Containers	备注 Remark
上海 Shanghai	中国 China	3 529	内外贸货物 Domestic and Foreign Trade Cargo
新加坡 Singapore	新加坡 Singapore	3 387	内外贸货物 Domestic and Foreign Trade Cargo
深圳 Shenzhen	中国 China	2 404	内外贸货物 Domestic and Foreign Trade Cargo
香港 Hong Kong	中国 China	2 223	内外贸货物 Domestic and Foreign Trade Cargo
宁波—舟山 Ningbo and Zhoushan	中国 China	1 945	内外贸货物 Domestic and Foreign Trade Cargo
釜山 Pusan	韩国 Korea, Rep.	1 865	内外贸货物 Domestic and Foreign Trade Cargo
青岛 Qingdao	中国 China	1 658	内外贸货物 Domestic and Foreign Trade Cargo
广州 Guangzhou	中国 China	1 639	内外贸货物 Domestic and Foreign Trade Cargo
迪拜 Dubayy	阿联酋 United Arab Em	1 525	内外贸货物 Domestic and Foreign Trade Cargo
天津 Tianjin	中国 China	1 406	内外贸货物 Domestic and Foreign Trade Cargo
鹿特丹 Rotterdam	荷兰 Netherlands	1 234	内外贸货物 Domestic and Foreign Trade Cargo
巴生 Kelang	马来西亚 Malaysia	1 094	内外贸货物 Domestic and Foreign Trade Cargo
高雄 Gaoxiong	中国台湾 Taiwan, China	1 059	内外贸货物 Domestic and Foreign Trade Cargo
大连 Dalian	中国 China	1 013	内外贸货物 Domestic and Foreign Trade Cargo
汉堡 Hamburg	德国 Germany	973	内外贸货物 Domestic and Foreign Trade Cargo
安特卫普 Antwerp	比利时 Belgium	896	内外贸货物 Domestic and Foreign Trade Cargo
厦门 Xiamen	中国 China	857	内外贸货物 Domestic and Foreign Trade Cargo
丹戎帕拉帕斯 Tanjung Periuk	马来西亚 Malaysia	852	内外贸货物 Domestic and Foreign Trade Cargo
洛杉矶 Los Angeles	美国 United States	834	内外贸货物 Domestic and Foreign Trade Cargo
长滩 Long Beach	美国 United States	682	内外贸货物 Domestic and Foreign Trade Cargo

资料来源：上海航运交易所。

Data Source: *Shanghai Shipping Exchange.*

11-19 港口货物吞吐量居世界前20位的港口（2013年）
World Top 20 Seaports in Terms of the Cargo Handled, 2013

单位：百万吨 (million t)

港口 Seaport	所属国家或地区 Country or Region	吞吐量 Cargo Handled	备注 Remark
上海 Shanghai	中国 China	775. 7	内外贸货物 Domestic and Foreign Trade Cargo
新加坡 Singapore	新加坡 Singapore	560. 9	内外贸货物 Domestic and Foreign Trade Cargo
天津 Tianjin	中国 China	500. 6	内外贸货物 Domestic and Foreign Trade Cargo
宁波 Ningbo	中国 China	495. 9	内外贸货物 Domestic and Foreign Trade Cargo
广州 Guangzhou	中国 China	455. 2	内外贸货物 Domestic and Foreign Trade Cargo
青岛 Qingdao	中国 China	450. 0	内外贸货物 Domestic and Foreign Trade Cargo
鹿特丹 Rotterdam	荷兰 Netherlands	440. 5	内外贸货物 Domestic and Foreign Trade Cargo
大连 Dalian	中国 China	407. 5	内外贸货物 Domestic and Foreign Trade Cargo
黑德兰港 Headland Harbour	澳大利亚 Australia	372. 3	内外贸货物 Domestic and Foreign Trade Cargo
釜山 Pusan	韩国 Korea, Rep.	313. 3	内外贸货物 Domestic and Foreign Trade Cargo
香港 Hong Kong	中国 China	276. 1	内外贸货物 Domestic and Foreign Trade Cargo
秦皇岛 Qinhuangdao	中国 China	272. 6	内外贸货物 Domestic and Foreign Trade Cargo
南路易斯安娜 Southern Louisiana	美国 United States	241. 6	外贸货物 Foreign Trade Cargo
休斯敦 Houston	美国 United States	236. 5	内外贸货物 Domestic and Foreign Trade Cargo
深圳 Shenzhen	中国 China	234. 0	内外贸货物 Domestic and Foreign Trade Cargo
名古屋 Nagoya	日本 Japan	208. 2	内外贸货物 Domestic and Foreign Trade Cargo
巴生 Kelang	马来西亚 Malaysia	198. 9	内外贸货物 Domestic and Foreign Trade Cargo
安特卫普 Antwerp	比利时 Belgium	190. 8	内外贸货物 Domestic and Foreign Trade Cargo
丹皮尔 Dampier	澳大利亚 Australia	177. 5	内外贸货物 Domestic and Foreign Trade Cargo
蔚山 Ulsan	韩国 Republic of Korea	167. 9	内外贸货物 domestic and foreign trade cargo

资料来源：Shipping Statistics and Market Review, December 2014, ISL.

Data Source: *Shipping Statistics and Market Review, December 2014, ISL.*

11-20　海上商船拥有量居世界前20位的国家或地区（2013年）
World Top 20 Countries or Regions in Terms of the Number of Maritime Merchant Ships Owned ,2013

国家和地区 Country and Region	艘数 （艘） Number of Vessels (unit)	总吨 Gross Ton （千吨） (1 000 tons)	载重吨 Deadweight ton	
			千吨 1 000 tons	占世界% Percentage in the World
世界总计 World Total	**49 576**	**1 065 202**	**1 604 374**	**100.0**
巴拿马 Panama	6 971	224 607	346 660	21.6
利比里亚 Liberia	3 025	128 478	200 764	12.5
马绍尔群岛 Marshall Islands	2 047	88 701	145 727	9.1
中国香港 Hong Kong, China	2 169	83 079	137 432	8.6
新加坡 Singapore	2 056	65 936	101 469	6.3
希腊 Greece	1 081	43 324	77 038	4.8
马耳他 Malta	1 669	46 336	72 483	4.5
中国 China	2 577	44 454	70 690	4.4
巴哈马 Bahamas	1 144	48 207	63 212	3.9
英国 United Kingdom	887	29 922	40 439	2.5
塞浦路斯 Cyprus	822	20 342	31 998	2.0
日本 Japan	2 536	15 141	19 977	1.2
意大利 Italy	777	17 393	19 569	1.2
挪威 Norway	808	14 819	18 884	1.2
韩国 Korea, Rep.	1 000	10 402	16 844	1.0
德国 Germany	405	14 430	16 257	1.0
丹麦 Denmark	452	12 208	14 578	0.9
印度 India	603	8 422	14 545	0.9
安提瓜和巴布亚 Antigua and Papua	1 178	10 277	13 301	0.8
印度尼西亚 Indonesia	2 483	9 059	12 613	0.8

注：统计截止日期为2014年1月1日。

资料来源：SSMR, Jan/Feb 2014, ISL (SSMR 2013年商船数据来源于 Clarkson Research Kservices Limited, CRSL).

Note: The closing date of statistics was 1 Jan., 2014.

Data Source: SSMR, Jan/Feb 2014, ISL. (The SSMR 2013 merchant ship data is derived from Clarkson Research Kservices Limited, CRSL).

11-21 集装箱船拥有量居世界前20位的国家或地区（2013年）
World Top 20 Countries or Regions in Terms of the Number of Container Ships Owned, 2013

国家和地区 Country and Region	艘数 （艘） Number of Vessels (unit)	载重吨 Deadweight ton	
		千标准箱 1 000 TEU	占世界% Percentage in the World
世界总计 World Total	**5 106**	**17 125**	**100.0**
利比里亚 Liberia	984	3 801	22.2
巴拿马 Panama	719	3 007	17.6
中国香港 Hong Kong, China	344	1 591	9.3
新加坡 Singapore	435	1 491	8.7
德国 Germany	245	1 166	6.8
英国 United Kingdom	162	869	5.1
马绍尔群岛 Marshall Islands	245	732	4.3
丹麦 Denmark	104	693	4.0
马耳他 Malta	134	528	3.1
安提瓜和巴布亚 Antigua and Papua	373	509	3.0
中国 China	192	465	2.7
塞浦路斯 Cyprus	203	413	2.4
美国 United States	68	226	1.3
法国 France	28	210	1.2
希腊 Greece	30	193	1.1
巴哈马 Bahamas	57	137	0.8
印度尼西亚 Indonesia	186	113	0.7
韩国 Korea, Rep.	88	89	0.5
荷兰 Netherlands	53	88	0.5
中国台湾省 Taiwan, China	27	88	0.5

注：1. 统计截止日期为2014年1月1日。

2. 因统计口径不同，表中数据与其他出版物公布的数据略有差别。

资料来源：SSMR May/June 2014, ISL（SSMR2013年集装箱船数据来源于Clarkson Research Services Limited, CRSL）.

Note: 1. The closing date of statistics was 1 Jan., 2014.

2. Owing to different statistical requirements, the data given in the table may be somewhat different from those issued in other publications.

Data Source: SSMR May/June 2014, ISL. (The SSMR 2013 container ship data is derived from Clarkson Research Services Limited, CRSL).

11-22 海上油轮拥有量居世界前20位的国家或地区（2013年）
Top 20 Countries or Regions in the World by the Possession of Ocean Going Oil Tankers, 2013

国家和地区 Country and Region	艘数（艘） Number of Vessels (unit)	总吨 Gross Ton（千吨）(1 000 tons)	载重吨 Deadweight ton 千吨 1 000 tons	载重吨 Deadweight ton 占世界% Percentage in the World
世界总计 World Total	**13 290**	**342 436**	**567 933**	**100.0**
利比里亚 Liberia	940	44 164	77 399	13.6
巴拿马 Panama	1 530	45 759	76 258	13.4
马绍尔群岛 Marshall Islands	836	42 092	69 128	12.2
希腊 Greece	454	25 684	47 318	8.3
新加坡 Singapore	990	24 375	42 096	7.4
巴哈马 Bahamas	387	25 403	40 248	7.1
中国香港 Hong Kong, China	424	16 582	29 277	5.2
马耳他 Malta	543	15 894	27 555	4.9
中国 China	645	8 463	14 358	2.5
英国 United Kingdom	290	8 917	14 188	2.5
坦桑尼亚 Tanzania	61	5 841	11 126	2.0
挪威 Norway	223	7 457	11 062	1.9
印度 India	122	4 851	8 662	1.5
意大利 Italy	267	5 028	8 363	1.5
日本 Japan	862	5 536	6 776	1.2
塞浦路斯 Cyprus	131	3 914	6 342	1.1
马来西亚 Malaysia	196	4 804	6 307	1.1
百慕大 Bermuda	71	5 974	6 203	1.1
丹麦 Denmark	163	3 466	5 443	1.0
印度尼西亚 Indonesia	509	2 982	4 760	0.8

注：统计截止日期为2014年1月1日。

资料来源：Total tanker fleet by ownership patterns, SSMR March 2014, ISL.

Note: The closing date of statistics was 1 Jan., 2014.

Data Source: Total tanker fleet by ownership patterns, SSMR March 2014, ISL.

11-23 海上散货船拥有量居世界前20位的国家或地区（2013年）

Top 20 Countries or Regions in the World by the Possession of Ocean Bulk Carriers, 2013

国家和地区 Country and Region	艘数（艘） Number of Vessels (unit)	总吨（千吨） (1 000 tons)	载重吨 Deadweight ton 千吨 1 000 tons	载重吨 Deadweight ton 占世界% Percentage in the World
世界总计 World Total	**10 381**	**388 053**	**705 826**	**100.0**
巴拿马 Panama	2 686	114 410	210 377	29.8
中国香港 Hong Kong, China	1 025	45 717	83 862	11.9
利比里亚 Liberia	809	38 276	70 154	9.9
马绍尔群岛 Marshall Islands	796	34 908	63 771	9.0
中国 China	968	26 577	45 722	6.5
新加坡 Singapore	383	19 431	35 971	5.1
马耳他 Malta	537	19 327	34 882	4.9
希腊 Greece	271	14 254	26 898	3.8
塞浦路斯 Cyprus	277	10 194	18 469	2.6
巴哈马 Bahamas	262	8 570	15 076	2.1
英国 United Kingdom	123	7 291	13 694	1.9
韩国 Korea, Rep.	182	6 436	11 957	1.7
日本 Japan	376	5 585	10 277	1.5
意大利 Italy	94	4 163	7 666	1.1
印度 India	95	2 683	4 808	0.7
土耳其 Turkey	100	2 587	4 467	0.6
挪威 Norway	65	2 475	4 370	0.6
菲律宾 Philippines	83	2 396	4 034	0.6
比利时 Belgium	22	1 712	3 307	0.5
百慕大 Bermuda	19	1 492	2 878	0.4

注：统计截止日期为2014年1月1日。

资料来源：Total bulk carrier fleet by ownership patterns, SSMR April 2014, ISL.

Note: The closing date of statistics was 1 Jan., 2014.

Data Source: Total bulk carrier fleet by ownership patterns, SSMR April 2014, ISL.

图书在版编目（CIP）数据

中国海洋统计年鉴. 2015 : 汉、英 / 国家海洋局编. —北京：海洋出版社，2016.6

ISBN 978-7-5027-9487-3

Ⅰ. ①中… Ⅱ. ①国… Ⅲ. ①海洋－统计资料－中国－2015－年鉴－汉、英 Ⅳ. ①P7-66

中国版本图书馆 CIP 数据核字(2016)第 122256 号

责任编辑：王　溪

责任印制：赵麟苏

海洋出版社 出版发行

http: //www.oceanpress.com.cn

北京市海淀区大慧寺路 8 号　　　　邮编：100081

国家海洋信息中心印刷厂印刷　　　新华书店发行所经销

2016 年 6 月第 1 版　　2016 年 6 月第 1 次印刷

开本：787mm×1092mm　1/16　印张：19.75

字数：495 千字

定价：168.00 元

发行部：(010)62147016　　邮购部：(010)68038093　　总编室：(010)62114335